Analog Circuits and Signal Processing

For further volumes:
http://www.springer.com/series/7381

Daniele Raiteri • Eugenio Cantatore
Arthur H.M. van Roermund

Circuit Design
on Plastic Foils

 Springer

Daniele Raiteri
Department of Electrical Engineering
Eindhoven University of Technology
Eindhoven
The Netherlands

Arthur H.M. van Roermund
Department of Electrical Engineering
Eindhoven University of Technology
Eindhoven
The Netherlands

Eugenio Cantatore
Department of Electrical Engineering
Eindhoven University of Technology
Eindhoven
The Netherlands

ISSN 1872-082X
ISBN 978-3-319-11426-2
DOI 10.1007/978-3-319-11427-9
Springer Cham Heidelberg New York Dordrecht London

ISSN 2197-1854 (electronic)
ISBN 978-3-319-11427-9 (eBook)

Library of Congress Control Number: 2014953546

Printed on acid-free paper

Springer is part of Springer Science+Business Media (www.springer.com)

Contents

List of Symbols

C_{dec} Decoupling capacitor [F]
C_i Insulator capacitance per unit area [F/μm^2]
C_L Load capacitance [F]
C_p Parasitic capacitance [F]
C_{SG} Source to gate capacitor [F]
C_{TGG} Top-gate to gate capacitor [F]
C_{TGS} Top-gate to source capacitor [F]
E_i Energy level of the ith localized state [J]
E_p Pinch-off electric field [V/cm]
f_t Cut-off frequency [Hz]
FW Finger Width [μm]
G_{diff} Differential gain [V/V]
G_m Transconductance of the operational amplifier [A/V]
g_m Transconductance of the transistor [A/V]
K_B Boltzmann constant [J/K]
L Channel length [μm]
L_G Length of the gate metal in the PC [μm]
N_t Number of localized/trap states
Q_D Channel charge at the drain side [C]
Q_S Channel charge at the source side [C]
R'_{sub} Substrate resistance [Ω]
R_L Load resistance [Ω]
SC Sub-Channels
T Working temperature of the transistor [K]
T_0 Semiconductor disorder characteristic temperature [K]
T_{CLK} Clock period [s]
V_D Drain voltage [V]
V_{DD} Supply voltage [V]
V_{DS} Drain-source voltage [V]
V_{FB} Flat-band voltage [V]
V_G Gate voltage [V]
V_{OD} Overdrive voltage [V]

$\mathbf{V}_p$	Equivalent Early voltage [V]
$\mathbf{V}_S$	Source voltage [V]
$\mathbf{V}_{SS}$	Sub-threshold slope [V]
$\mathbf{V}_{TG}$	Top-gate Voltage [V]
$\mathbf{W}_G$	Width of the gate metal in the PC [μm]
$\mathbf{W}_{OSC}$	Width of the OSC in the PC [μm]
Γ	Traps coefficient [K/K]
ε_0	Electric constant [F/m]
ε_r	Dielectric constant
$\mathbf{H}$	Top-gate coupling parameter [μm/μm]
θ	Heaviside unit step function

List of Abbreviations

AC	Alternating Current
AD	Analog to Digital
ADC	Analog to Digital Converter
ALU	Arithmetic Logic Unit
AMOLED	Active-Matrix Organic Light-Emitting Diode
CB	Conduction Band
CMOS	Complementary Metal Oxide Semiconductor
CPU	Central Processing Unit
CS	Current Steering
DA	Digital to Analog
DC	Direct Current
DFF	Data Flip Flop
DNL	Differential Non-Linearity
DOS	Density of States
DPCA	Differential Parametric Capacitor Amplifier
DR	Dynamic Range
DRC	Design Rule Checker
EDA	Electronic Design Automation
ENOB	Effective Number of Bits
FW	Finger Width
GBW	Gain BandWidth
GIZO	Gallium-Indium-Zinc-Oxide
HD	High Definition
IC	Integrated Circuit
INL	Integral Non-Linearity
IPMC	Ionic Polymer Metal Composite
LED	Light-Emitting Diode
LS	Level Shifter
LSB	Least Significant Bit
LTPS	Low Temperature PolySilicon
LVS	Layout vs Schematic
MEC	Maximum Equal Criterion

MOSFET	Metal-Oxide-Semiconductor Field Effect Transistor
MSB	Most Significant Bit
MTR	Multiple Trapping and Release
NM	Noise Margin
OC	Output Characteristic
OSC	Organic SemiConductor
OTA	Operational Transconductance Amplifier
PBG	Planar Bottom-Gate
PC	Parametric Capacitor
PCell	Parameterized Cell
PLS	Positive-feedback Level Shifter
PTG	Planar Top-Gate
RF	Radio Frequency
RFID	Radio Frequency IDentification
SAM	Self-Assembled Monolayer
SAR	Successive Approximation Register
SBG	Staggered Bottom-Gate
SFDR	Spurious Free Dynamic Range
SNDR	Signal to Noise and Distortion Ratio
SNR	Signal to Noise Ratio
STG	Staggered Top-Gate
TC	Transfer Characteristic
TCTG	Top-gate Transfer Characteristic
TFT	Thin-Film Transistor
VCO	Voltage Controlled Oscillator
VRH	Variable Range Hopping

Chapter 1
Introduction

The first integrated circuits date back to the 1950s. From that moment, humanity witnessed an impressive growth of the presence of electronics in everyday life. This growth is not going to stop, and will also involve innovative applications that, contrary to most product nowadays, will not be based on high computational power or ultra-fast mobile data communication.

Many of these new applications will indeed demand e.g. large-area, mechanical flexibility, fast prototyping, customization and ultra-low-cost production. Examples include surfaces allowing human-machine interaction (tactile, visual, etc.) and bringing internet connection even to very cheap objects (thus widening the concept of "internet of things" to everything around us).

New electronic technologies, devices, circuits and systems are being created to empower this future. A first glimpse at this emerging electronics is given in this chapter, and a deeper insight will be provided in this whole book.

1.1 Silicon and Plastic

In 1965, Moore reported in his famous paper [1] that the number of transistors on an integrated circuit between 1958 and 1965 doubled every two years. He predicted the same trend to continue for at least ten years after his paper but his prediction turned out to be pessimistic. As a matter of fact, the integrated circuit industry adopted the well-known law named after him as the roadmap for its research and development, therefore Moore's trend has been followed till our days.

The exponential growth in the number of transistors per IC, fuelled by an impressive effort in device miniaturization and by outstanding achievements in the manufacturing processing, is tightly coupled to an exponential decrease of the unity cost of the transistor, which has enabled a pervasive presence of electronics in human lives. The downscaling of the feature size of silicon technologies does not allow only more integration, but also higher speed, reduced power-delay product, and improved reliability of the integrated circuits.

© Springer International Publishing Switzerland 2015
D. Raiteri et al., *Circuit Design on Plastic Foils,* Analog Circuits and Signal Processing,
DOI 10.1007/978-3-319-11427-9_1

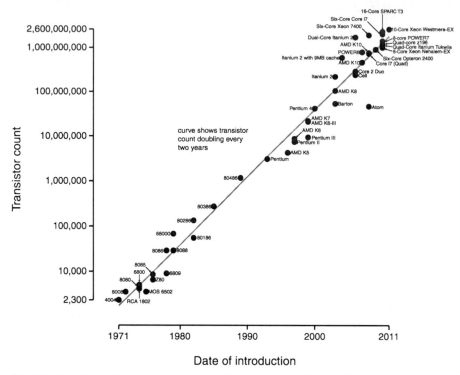

Fig. 1.1 Moore's law in microprocessor: transistor count versus history [2]

A classic example of Moore's law development can be found in the microprocessors, since the first Intel's i4004 in 1971. That first central processing unit (CPU) was built with a 4 bit architecture, it was manufactured with the first self-aligned technologies, and could work at a clock frequency of 740 kHz. Nowadays multi-core CPUs based on 64 bit architectures are on the market, they integrate graphic processors, and their clock runs at multi-GHz frequencies. As shown in Fig. 1.1, the number of transistor per IC has been growing exponentially in time in line with Moore's prediction, thanks to the impressive progress in the manufacturing process. Indeed the i4004 used a pioneering 10 μm minimum feature size self-aligned technology for its 2300 transistors; while nowadays the most advanced lithography processes for microprocessors reach 22 nm minimum feature size and these ICs count up to 2.6 billion transistors.

The ever growing computational power we witness fuels plenty of applications, like HD compact cameras, digital television, personal computers, laptops, game consoles, memories, tablets and smartphones. All these applications provide more and more functionalities even in portable devices, and achieve these results at a constant or even decreasing cost, thanks to Moore's development.

On the other hand, there are many applications of electronics that do not require high computational power, and cannot be addressed by standard IC technologies due to the high cost per area and to mechanical limitations. The typical example of such applications is a flat-screen display. Displays are intrinsically large-area. Their remarkable growth and irresistible commercial success are based on the development of a specific electronic technology, TFTs (Thin-Film Transistors) on glass, which aims at decreasing the cost per unit area (and not the cost per transistor, as it is the case in IC technology). TFTs are transistors manufactured on large glass carriers using thin semiconductor films deposited on the glass surface (amorphous or poly-crystalline silicon, in most cases).

In the last 15 years a clear trend towards lowering the processing temperature of large-area TFT processes has also been emerging. Lower temperatures indeed enable the use of thin, flexible plastic substrates that are much more attractive than the fragile and expensive glass sheets used nowadays. These developments have been fuelled by research on semiconductor materials that can be processed at near-to-ambient temperature, like organic and metal oxide semiconductors [3, 4].

Large-area electronics processed at low temperature on flexible substrates enable two sorts of innovative applications: on the one hand, applications that exploit together large area, flexibility and ruggedness, like touch screens, Braille displays [5], pedometers [6], strain gauges, artificial skin [7, 8] and more [9, 10]; on the other hand, applications that exploit these technologies for their simplicity and high throughput, aiming at achieving cost competitiveness with silicon for very cost-sensitive applications with low computational intensity. To this latter category belong, for instance, item-level Radio-Frequency Identification (RFID) tags [11] augmented with sensors, able to monitor the storage and distribution chain of food, pharmaceuticals or expensive chemicals.

Unfortunately large-area electronic processes on foil are still in an initial state of development, and the design of circuitry exploiting this technology must cope today with several drawbacks in terms of variability, ageing and low-performance. In this book, large-area technologies will be used to design building blocks suitable for smart sensors (which integrate electronics along with the actual sensor), with particular focus on the techniques adopted to face the several limitations that still affect the manufacturing process and the transistors on foil.

1.2   Challenges in Electronics on Flexible Substrates

Processing TFTs on large area at near-to-ambient temperatures is normally associated with poor control on the thickness of the single layers, on the spot temperature during the process, and other boundary conditions. For this reason, the variability among different TFTs is very pronounced, and it is difficult to match devices even if they are identical and close-by [12].

The variability and poor matching on the foil are not only due to process parameters varying from point to point, but also to the solid state structure of semiconductors suitable for low temperature processing. These materials are indeed normally dis-ordered micro or nano-crystalline materials, rather than perfectly uniform mono-crystals like in Silicon ICs. Another source of variability is the contact resistance between source/drain contacts and the semiconductor film [13, 14].

The intrinsic poor performance of semiconducting films processed at low tem-perature is closely related to their disordered packing. Disorder in the periodicity of the lattice prevents the charge transport from taking place in extended states; on the contrary charge carriers are in localized energy states [15] or "hop" [16] among them. For this reason, mobility typically ranges between 0.1 and 10 cm^2/V s [17], which is two or three orders of magnitude smaller than in crystalline silicon.

The reduced mobility translates in small TFT transconductance, while the output resistance is strongly limited by short-channel effects [18–20]. Indeed, in standard thin-film transistor technologies on flexible foils, the gate insulator cannot be scaled down to very thin layers and the control of the gate on the channel is heavily af-fected by the longitudinal electric field. These two aspects of large-area TFTs result in a maximum intrinsic gain typically ranging between 20 and 26 dB.

All these intrinsic TFT limitations are not the only issues concerning low-tem-perature semiconductor materials and their process technology. Indeed another im-portant aspect of large-area TFT technologies on foil is the availability of p-type, n-type or complementary transistors. Traditionally, low-temperature processes have been using only p-type organic semiconductors and thus p-type TFTs only [3]. Later on, also n-type TFTs have been processed at low temperature using organic materi-als [21] or metal oxides (like ZnO [15] or InGaZnO [22]). Even though metal-oxide semiconductors are rapidly becoming a standard technology in display backplane [23], their use in circuits is still in its early infancy.

Only recently, complementary processes have been demonstrated [24, 25] com-bining two organic materials or exploiting "hybrid technologies" that combine organic p-type with inorganic n-type semiconductors [26]. These processes are inherently more complex than unipolar ones, and typically result in higher defec-tivity. The most reliable (low defectivity and variability) TFT technologies fabri-cated at low temperature nowadays still offer, at the state of the art, only p-type normally-on organic TFTs and capacitors. State of the art p-type-only organic TFT circuits have reached a complexity level of 4000 transistors [27], while printed complementary organic technologies offer a maximum complexity of about 100 TFTs in functional circuits [28].

P-type TFTs in low-temperature technologies are typically normally-on devices. Semiconductors with such property are preferred since higher gain, better noise margins and thus higher circuit complexity can be achieved using a logic gate when the active load is implemented with a Zero-Vgs connected load (i.e. shorting gate and source of the TFT), compared to the diode-connected load which is the only vi-able solution for normally-off TFTs. However, normally-on TFTs represent a poor solution for switches or pass-transistors and do not necessarily work in saturation when diode-connected. Therefore, the design of current mirrors is not trivial on one supply rail (V_{DD} for p-type and gnd for n-type), and it is impossible at all on the other. The availability of only one type TFT poses strong limitations to the freedom of design for analog and digital circuits, and makes the design of reliable digital functions extremely challenging.

A final important issue, which has been only partially solved till now with material research, is the sensitivity of large-area low-temperature TFTs to aging and bias stress in air.

1.3 Book Aims and Overview

The aim of this book is to investigate the diverse major issues relevant to analog, digital and mixed-signal circuit design in large-area TFT technologies. Our first steps move from the modeling and the characterization of the devices and their implementation in a Computer Aided Design (CAD) environment. A clear understanding of the limitations posed by the technology is paramount to developed technology aware solutions applied at different abstraction levels: architectural, circuital, layout, or device.

Facing each issue at the right abstraction level allows pursuing the optimal solution which is not the one simply fulfilling all the electronic specifications required by the target application, but the one that guarantees high yield and good robustness to enable ultra-low-cost even for low-volume production. Also an increased technology complexity could represent a problem, as it invariably will translate into higher production costs. It is thus important to achieve as much as possible exploiting only unipolar technologies, which will always be cheaper than the complementary counterpart.

The work presented in this book focuses on circuit and system solutions for smart sensors manufactured using low-temperature large-area TFTs. The "smart sensor" is herewith defined as a system capable of physical measurements, first analog signal conditioning, data conversion and wireless data communication to a base station (Fig. 1.2).

Among the required building blocks, the sensor and the RF transceiver fall out of the scope of this book, while circuits necessary to the sensor frontend, as it is

Smart sensor

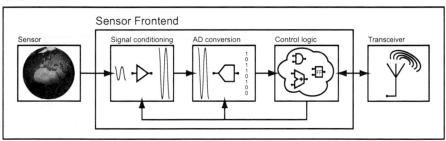

Fig. 1.2 Smart sensor schematic

defined here, have been studied and compared to the literature. More specifically, this book discuss circuits for analog signal conditioning, circuits for data conversion, and a new logic style to implement all the required control logic.

The main design goal is to improve the performance of analog, mixed-signal and digital circuits, and to increase their robustness against process variations and mismatch. In particular, we pursued the improvement of AD and DA converters' accuracy and the robustness of the digital logic.

In order to design the aforementioned sensor front-ends, two different technologies have been employed: the first is an organic double-gate p-type-only technology exploiting pentacene as semiconductor; the second one is a metal-oxide n-type-only technology.

Since the circuit complexity of analog, digital and mixed-signal blocks made with emerging technologies can appear trivial to the eye of a silicon IC designer, the book begins introducing, with Chaps. 2 and 3, large-area applications and technologies, and the state of the art in circuit design for these technologies. The comparison with standard electronics based on silicon clarifies the advantages and the limitations of both. Highlighting these differences is fundamental to understand the challenges imposed by low-cost low-temperature high-throughput manufacturing.

The electrical characterization and modeling of a double-gate p-type TFTs is presented in Chap. 4 with general remarks on process variations, mismatch, bias stress and ageing.

Chapter 5 illustrates the main building blocks necessary to build a smart sensor frontend and the architectural choices made for each taking into account features and drawbacks of the technology.

Chapters 6–8 deal with the design of the sensor frontend at a circuit level: from the analog signal conditioning, through the analog to digital data conversion, to the design of a new logic style suitable for robust digital processing in unipolar technologies.

Finally, conclusions are presented in Chap. 9.

References

1. G.E. Moore. Cramming more components onto integrated circuits. Electronics **38**, 114–117 (1965)
2. [Online] Moore's law, http://en.wikipedia.org/wiki/Moore's_law
3. G.H. Gelinck, T.C.T. Geuns, D.M. de Leeuw, High-pergformance all-polymer integrated circuits. Appl. Phys. Lett. **77**(10), 1487–1489 (2000)
4. A.K. Tripathi et al., Low-voltage gallium-indium-zinc-oxide thin film transistors based logic circuits on thin plstic foil: Building blocks for radio frequency identification application. Appl. Phys. Lett. **98**(16), 162102 (2011)
5. Y. Kato et al., Sheet-type Braille displays by integrating organic field-effect transistors and polymeric actuators. IEEE Trans. Electron Devices **54**(2), 202–209 (2007)
6. K. Ishida et al., Insole pedometer with piezoelectric energy harvester and 2 V organic circuits. IEEE J. Solid-State Circuits **48**(1), 255–264 (2013)
7. P. Cosseddu et al., Inkjet printed organic thin film transistors based tactile transducers for artificial robotic skin, in *IEEE RAS and EMBS International Conference on Biomedical Robotics and Biomechatronics*, pp. 1907–1912, 2012

8. T. Hoshi, H. Shinoda, Robot skin based on touch-area-sensitive tactile element, in *IEEE International Conference on Robotics and Automation,* pp. 3463–3468, 2006

9. K. Ishida et al., A 100-V AC energy meter integrating 20 V organic CMOS digital and analog circuits with a floating gate for process variation compensation and a 100 V organic pMOS rectifier. IEEE J. Solid-State Circuits **47**(1), 301–309 (2012)

10. T. Yokota et al., Sheet-type flexible organic active matrix amplifier system using pseudo-CMOS circuits with floating-gate structure. IEEE Trans. Electron Devices **59**(12), 3434–3441 (2012)

11. E. Cantatore et al., A 13.56 MHz RFID system based on organic transponders. IEEE J. Solid-State Circuits **42**(1), 82–92 (2007)

12. H. Fuketa et al., 1 um-thickness 64-channel surface electromyogram measurement sheet with 2 V organic transistors for prosthetic hand control, in *IEEE International Solid-State Circuits Conference*, San Francisco, pp. 104–105, 2013

13. P.V. Pesavento, K.P. Puntambekar, C.D. Frisbie, J.C. McKeen, P.P. Ruden, Film and contact resistance in pentacene thin-film transistors: dependence on film thickness, electrode geometry, and correlation with hole mobility. J. Appl. Phys. **99**(9), 094504 (2006)

14. R.J. Chesterfield, Variable temperature film and contact resistance measurements on operating n-channel organic thin film transistors. J. Appl. Phys. **95**(11), 6396 (2004)

15. F. Torricelli, J. Meijboom, Transport physics and device modeling of zinc oxide thin-film transistors Part I: Long-channel devices. IEEE Trans. Electron Devices **58**(8), 2610–2619 (2011)

16. M. Vissenberg, M. Matters, Theory of the field-effect mobility in amorphous organic transistors. Phys. Rev. B **57**(20), 964–967 (1998)

17. C.D. Dimitrakopoulos, P.R.L. Malenfant, Organic thin film transistors for large area electronics. Adv. Mater. **14**(2), 99–117 (2002)

18. N.H. Touidjen, F. Mansour, Modeling kink effect in the poly-Si TFTs under charge sheet approach, in *International Conference on Advances in Computational Tools for Engineering Applications*, pp. 407–410, 2009

19. L. Mariucci et al., Polysilicon TFT structures for kink-effect suppression. IEEE Trans. Electron Devices **51**(7), 1135–1142 (2004)

20. C.R. Wie, Nonsaturating drain current characteristic in short-channel amorphous-silicon thin-film transistors. IEEE Trans. Electron Devices **57**(4), 846–854 (2010)

21. A. Facchetti et al., Organic-inorganic flexible and transparent electronics, in *Flexible Electronics and Displays Conference and Exhibition*, pp. 1–7, 2008

22. H. Hosono, M. Yasukawa, H. Kawazoe, Novel oxide amorphous semiconductors: transparent conducting amorphous oxides. J. Non-Cryst. Solids **203**, 334–344 (1996)

23. L.G. 55-inch OLED TV, [Online], http://reviews.cnet.co.uk/tvs/lg-55-inch-oled-tv-review-50006604/

24. T. Zaki et al., A 3.3 V 6-bit 100 kS/s current-steering digital-to-analog converter using organic p-type thin-film transistors on glass. IEEE J. Solid-State Circuits **47**(1), 292–300 (2012)

25. S. Jacob et al., High performance printed N and P-type OTFTs for complementary circuits on plastic substrate, in *IEEE European Solid-State Device Research Conference*, pp. 173–176, 2012

26. K. Myny et al., Bidirectional communication in an HF hybrid organic/solution-processed metal-oxide RFID tag, in *IEEE International Solid-State Circuits Conference*, pp. 312–314, 2012

27. K. Myny et al., An 8-bit, 40-instructions-per-second organic microprocessor on plastic foil. IEEE J. Solid-State Circuits **47**(1), 284–291 (2012)

28. S. Abdinia et al., A 4b ADC manufactured in a fully-printed organic complementary technology including resistors, in *IEEE International Solid-State Circuits Conference*, pp. 106–107, 2013

Chapter 2
Applications of Large-area Electronics on Foil

Applications addressed by large-area electronics give the motivation to push further the research on large-area TFT manufacturing processes and circuit design. Since the birth of electronics, the main research goals have been miniaturization and increasing computational power. Today's focus is turning back to people, and electronics aims now to pervade everyday life. From this prospective, standard transistor technologies do not provide anymore a solution suitable to all demands. For innovative applications mechanical properties may become more important than performance, or cost must be sometimes evaluated with respect to the area and not to the complexity of the function realized. In this chapter, several applications enabled by large area electronics on foil will be reviewed.

2.1 From Miniaturization to Flexible Substrates

The first integrated circuits date back to the 1950s. Since that moment, the miniaturization of the metal-oxide-semiconductor field effect transistors (MOSFETs) has enabled enough computational power for lots of applications ranging from medicine to gaming, from aviation to video editing, from automotive to telecommunications, and many others.

Nevertheless, more recently, different applications have also attracted the attention of academic and industrial researchers. In this case, the key features are represented by the mechanical properties of the electronics (large area and flexibility) and by the low-cost of the manufacturing process.

For this reason, new materials and deposition techniques suitable for cheap substrates, like paper, plastic foils and fabrics [1], have been investigated, which can be processed in ambient environment with potentially high throughput. Different technologies have been developed to realize a wide set of devices such as physical sensors [2–4], photo detectors [5], light emitters, plastic actuators [6] and general purpose flexible electronics [7]. Not all these components can be integrated on the same substrate and manufactured within the same process yet, but they share common

© Springer International Publishing Switzerland 2015
D. Raiteri et al., *Circuit Design on Plastic Foils,* Analog Circuits and Signal Processing,
DOI 10.1007/978-3-319-11427-9_2

manufacturing techniques and materials, hence the way to a full integration can be discerned at the horizon.

Of course, these new technologies do not aim to replace silicon ICs in high performance applications. Indeed, due to the low temperature nature of the process and to the use of materials suitable for deposition from solution, the electrical performance of large-area electronics is typically much lower than standard silicon technologies. The mobility of organic and metal-oxide semiconductors is usually about three orders smaller than that in silicon; the intrinsic gain of these transistors is a few tens; often only p- or n-type device are available; TFTs typically suffer from poor uniformity, bias stress and aging, which affect the parameters of thin-film transistors increasing the deviation from their nominal values.

2.2 Applications

Applications exploiting cheap flexible substrates, and cheap manufacturing process, can be divided at first glance in two main groups: large-area applications and low-cost applications. The first one aims to a reduction of the cost per area, while the second one to a reduction of the cost per transistor/function.

2.2.1 Large-area Applications

In the last two decades, the major force driving the innovation around large-area electronics was the industry of flat-panel displays. For this sort of applications indeed, electronics need to be manufactured on a surface much larger than common semiconductor wafers in order to drive all the pixels composing the display. In this domain the most interesting technology from a commercial viewpoint is thus the one achieving lower cost per area, not the one achieving lower cost per transistor (Moore's law).

At the beginning, TFT technologies developed for this purpose used to exploit glass substrates and polycrystalline or amorphous silicon [8, 9] as semiconductors. When materials suitable for low-temperature deposition enabled the integration of circuits on foils, lots of new applications became possible.

In the sphere of display applications, technologies on flexible substrates can be used for portable devices due to their higher ruggedness compared to the standard silicon ones. For example, if the display backplane is manufactured on flexible substrates, the touch screen would not crack if the portable device is accidentally dropped on the floor or if someone seats on it. Also, plastic foils are much lighter than glass substrates: a very convenient feature not only for wide screen televisions, but also for music players, tablets, e-book readers and smartphones. One of the most successful combinations of flexible electronics for display backplanes with new light-emitting materials can be found in Active-Matrix Organic Light-Emitting

Diode (AMOLED) displays which can be used in bendable, rollable and foldable portable systems, achieving at the same time low cost, large area, high resolution and low power consumption [10].

The interaction between light and charge carriers in semiconductors can also be exploited in flexible lighting surfaces [11] and solar cells [12]. In the first case, the semiconductor is used to convert electrical power in photon emission, while in the second it converts light into electric current. The use of large-area technologies for these applications would considerably decrease the cost of materials, production and shipment. Solar panels manufactured with these technologies are lighter and rugged; moreover the disposal of old cells would be easier, safer and thus cheaper.

The integration of electronics with plastic polymers enables actuating surfaces as Braille displays [6]. A matrix of organic transistors, similar to a display backplane, is used for pixel addressing. A dielectric elastomer can be bent applying a certain bias voltage to create a Braille symbol. When the ionic polymer-metal composite (IPMC) strip is properly biased, the plastic hemisphere connected to its untied end is pushed against the top surface and can thus be read by the blind person.

Large-area electronics can also be employed together with many kinds of surface sensors, to enable sensing surfaces, like sheet-sensors [13], pocket scanners [14], and others [6, 13, 15–18]. For this sort of applications, mechanical flexibility plays a much more important role than TFT performance. For instance, surface pressure sensors on a bendable plastic foil that can be curved up to a radius of 2 cm were used to realize the first e-skin, providing tactile inputs to a human sized robot hand [19, 20]. Organic light sensors have also been used to realize a portable black & white scanner, where the matrix of light sensors along with their readout transistors was manufactured on a transparent bendable plastic foil. The light shining on the sheet-type scanner can either be reflected by the bright surface or absorbed by the dark colors. Functional peripheral organic ADCs (Analog to Digital Converters) could then be used to convert the current generated by the sensors in a grayscale digital picture.

All these applications give only a flavor of the potential use of large-area electronics on flexible foils. They show indeed how large-area electronics can interact in many different ways with both nature and human beings. It can sense different quantities, transmit electrical signals and irradiate light, and even provide mechanical actuation through dielectric elastomers.

2.2.2 Low-Cost Applications

Exploiting technologies suitable for large area, many kinds of sensors have been demonstrated, also beyond the domain of smart surfaces. Temperature, strain, chemical, and biomedical sensors [21] can also be used for ultra-low-cost applications like circuits integrated in packaging at item level. A unique process integrating sensors and electronics paves the way towards disposable smart sensors: systems aiming at the detection of a physical quantity and able to perform simple actions when required.

In these applications the electronic function becomes more important, since many building blocks needs to be designed in order to process the sensor analog signal and make it amenable for post processing. Analog, digital and mixed-signal circuits must be designed based on technologies that are often superficially characterized, affected by large mismatch and process variations, and unstable under bias and environmental aggressions. Once these issues will be overcome with the help of circuit and technology solutions, we will witness e.g. smart sensors integrated in the package of food or pharmaceuticals to test their conservation quality, assays to monitor the level of bacteria in water, and robots equipped with conformable electronic skin [19] sensing roughness, softness, temperature and other qualities of our world.

Let us consider a smart sensor in the package of food. Every single item in a refrigerating room, or in the trailer of a truck, could detect locally its own temperature. This information could be used for instance to notify the final customer either if the desired product suffered multiple defrosts, or if it was perfectly stored and freshly delivered to the grocery shelf.

A chemical sensor integrated in a can of tomato sauce could reveal an excessive acidity and switch on an expiration LED. Or, in pharmaceuticals, it could monitor some excessive chemical levels in the medicine and communicate the automated storing system to withdraw the item from the market.

In order to perform such activities, the simplest smart sensor only requires a few building blocks (Fig. 1.2): a sensor, the sensor frontend, and a link to the base station that can perform more complex data processing and control more elaborated actions. In the following, the "sensor frontend" will be investigated.

The frontend chain begins with an analog signal conditioning interface: first, to reduce the impact of the noise introduced by the electronics on the weak input signal; second, to amplify the analog signal to a voltage range compatible with the data conversion. These operations are not trivial since the supply voltage of large-area electronics can reach several tens of volts and the intrinsic gain of the transistors is usually low. Moreover the lack of a complementary technology prevents the use of almost all the well-known architectures used in CMOS technology for operational amplifier designs. Indeed, the low intrinsic gain of TFTs makes negative-feedback systems design very challenging, hampering the effectiveness of linearity-enhancement and offset-reduction techniques.

The second step is the analog-to-digital conversion, required to achieve a robust data transmission to the base station. Also in this case, the poor performance of the electronics poses severe hurdles for the design of analog-to-digital and digital-to-analog converters. Even between closely-placed transistors, the process parameter variations and the large TFT mismatch reflect in significant offset at the input of e.g. comparators and OpAmps, hampering their use in ADC topologies like flash or pipelines. Only ADC architectures which are more resilient to comparator offset and exploit the relatively better matching of passives like SARs have been shown in literature, but even in this case the maximum linearity achieved was of about 5 bit [22]. When following oversampling and noise-shaping approaches, the designer has to deal with the low cut-off frequency of the organic transistors, which reflects

RAITERI, DANIELE.

CIRCUIT DESIGN ON PLASTIC FOILS.

Cloth   130 P.

CHAM: SPRINGER, 2015
SER: ANALOG CIRCUITS AND SIGNAL PROCESSING
SERIES.
AUTH: EINDHOVEN UNIVERSITY OF TECHNOLOGY.

LCCN 2014953546
  ISBN 3319114263   **Library PO#** SLIP ORDERS

		List	99.99	USD
6207 UNIV OF TEXAS/SAN ANTONIO	**Disc**	17.0%		
App. Date 8/26/15 EEN.APR 6108-09	**Net**	82.99	USD	

SUBJ: ELECTRONIC CIRCUIT DESIGN.

CLASS TK7867   DEWEY# 621.4   LEVEL ADV-AC

YBP Library Services

RAITERI, DANIELE.

CIRCUIT DESIGN ON PLASTIC FOILS.

Cloth   130 P.

CHAM: SPRINGER, 2015
SER: ANALOG CIRCUITS AND SIGNAL PROCESSING
SERIES.
AUTH: EINDHOVEN UNIVERSITY OF TECHNOLOGY.

LCCN 2014953546
  ISBN 3319114263   **Library PO#** SLIP ORDERS

		List	99.99	USD
6207 UNIV OF TEXAS/SAN ANTONIO	**Disc**	17.0%		
App. Date 8/26/15 EEN.APR 6108-09	**Net**	82.99	USD	

SUBJ: ELECTRONIC CIRCUIT DESIGN.

CLASS TK7867   DEWEY# 621.4   LEVEL ADV-AC

on a limited gain-bandwidth product in the OpAmps, so that the linearity achieved at the state of the art with these approaches is even lower [23].

Digital building blocks are also required in order e.g. to control the data flow, to restore signal synchronization, or to assist the data conversion. In this case, the limited static performance of the TFTs and the lack of complementary transistors affect the robustness of the digital circuits, resulting in a relatively high chance of non-functional circuits due to both hard and soft faults (i.e. errors due respectively to hardware faults like shorts and lines stuck at V_{DD} or gnd, or simply to excessive parameter variations). For this reason, the design of a robust digital logic is also mandatory.

Nowadays unipolar technologies are by far the most reliable in terms of hard faults and, for this reason, they provide the best solution for complex circuit design. However, with increasing process reliability, also complementary technologies will eventually offer similar hard-yield. At that point, soft faults will limit the maximum circuit complexity and unipolar technologies will need smart solutions to achieve yield levels comparable to the ones that can be achieved with complementary ones. This may still be interesting for practical applications, because of the far cheaper process that unipolar technologies enable.

References

1. W.S. Wong, A. Salleo, *Flexible Electronics: Materials and Applications* (Springer, New York, 2009)
2. S. Jung, T. Ji, V.K. Varadan, Temperature sensor using thermal transport properties in the subthreshold regime of an organic thin film transistor. Appl. Phys. Lett. **90**(6), 062105 (2007)
3. Y. Noguchi, T. Sekitani, T. Someya, Organic-transistor-based flexible pressure sensors using ink-jet-printed electrodes and gate dielectric layers. Appl. Phys. Lett. **89**(25), 253507 (2006)
4. M.D. Angione et al., Carbon based materials for electronic bio-sensing. Mater. Today **14**(9), 424–433 (2011)
5. I. Nausieda et al., An organic active-matrix imager. IEEE Trans. Electron Devices **55**(2), 527–532 (2008)
6. Y. Kato et al., Sheet-type Braille displays by integrating organic field-effect transistors and polymeric actuators. IEEE Trans. Electron Devices **54**(2), 202–209 (2007)
7. K. Ishida et al., User customizable logic paper (UCLP) with sea-of transmission-gates (SOTG) of 2-V organic CMOS and ink-jet printed interconnects. IEEE J Solid-State Circuits **46**(1), 285–292 (2011)
8. M. Stewart et al., Polysilicon VGA active matrix OLED displays-technology and performance, in *International Electron Devices Meeting,* pp. 871–874, 1998
9. J.K. Lee et al., a-Si:H thin-film transistor-driven flexible color e-paper display on flexible substrates. IEEE Electron Device Lett. **31**(8), 833–835 (2010)
10. V. Vaidya, D.M. Wilson, X. Zhang, B. Kippelen, An organic complementary differential amplifier for flexible AMOLED applications, in *IEEE International Symposium on Circuits and Systems,* pp. 3260–3263, 2010
11. Flexible Organic LED (OLED) lighting reaches high energy efficiency (2013). http://www.holstcentre.com/en/NewsPress/PressList/Solvay_OLED_Eng.aspx. Last accessed 26 Oct 2014
12. Y. Galagan, et al., Technology development for roll-to-roll production of organic photovoltaics. Chem. Eng. Process. **50**(5–6), 454–461 (2011)

13. T. Yokota, et al., Sheet-type flexible organic active matrix amplifier system using pseudo-CMOS circuits with floating-gate structure. IEEE Trans. Electron Devices **59**(12), 3434–3441 (2012)
14. T. Someya, et al., Pocket scanner using organic transistors and detectors, in IEEE *Lasers and Electro-Optics Society Annual Meeting,* pp. 76–77, 2005
15. K. Ishida, et al., Insole pedometer with piezoelectric energy harvester and 2 V organic circuits. IEEE J. Solid-State Circuits **48**(1), 255–264 (2013)
16. K. Ishida, et al., A 100-V AC energy meter integrating 20 V organic CMOS digital and analog circuits with a floating gate for process variation compensation and a 100 V organic pMOS rectifier. IEEE J. Solid-State Circuits **47**(1), 301–309 (2012)
17. T. Sekitani, et al., Stretchable active-matrix organic light-emitting diode display using printable elastic conductors. Nat. Mater. **8**(6), 494–499 (2009)
18. T. Sekitani, et al., Communication sheets using printed organic nonvolatile memories, in *IEEE International Electron Devices Meeting,* pp. 221–224, 2007
19. P. Cosseddu, et al., Inkjet printed organic thin film transistors based tactile transducers for artificial robotic skin, in *IEEE RAS and EMBS International Conference on Biomedical Robotics and Biomechatronics,* pp. 1907–1912, 2012
20. T. Hoshi, H. Shinoda, Robot skin based on touch-area-sensitive tactile element, in *IEEE International Conference on Robotics and Automation,* pp. 3463–3468, 2006
21. R.P. Singh, J.-W. Choi, Biosensors development based on potential target of conducting polymers. Sens. Transducers J. **104**(5), pp. 1–18 (2009)
22. W. Xiong, Y. Guo, U. Zschieschang, H. Klauk, B. Murmann, A 3-V, 6-Bit C-2C digital-to-analog converter using complementary organic thin-film transistors on glass. IEEE. Solid-State Circuits **45**(7), 1380–1388 (2010)
23. H. Marien, M.S.J. Steyaert, E. van Veenendaal, P. Heremans, A fully integrated ΔΣ ADC in organic thin-film transistor technology on flexible plastic foil. IEEE J. Solid-State Circuits **46**(1), 276–284 (2011)

Chapter 3
State of the Art in Circuit Design

The technologies that have been developed to address large-area and low-cost applications have been developed very recently and still pose severe limitations to the complexity achievable by integrated circuits. The transistor performance is limited and the process variations still represent a concrete issue with respect to circuit yield. In this chapter, the state of the art of large-area technologies is described, highlighting the main differences with respect to the most common standard silicon technologies. An overview is provided on the effort spent to build up a design flow platform easily configurable to address different and future technologies. The chapter concludes summarizing the state of the art in circuit design, which actually demonstrates all the difficulties that are experienced in this field when exploiting these technologies.

3.1 Introduction

Providing a complete survey of all different large-area technologies is a challenging task. The variety of substrates and semiconductors, deposition techniques and patterning processes is almost countless. Anyway, the most relevant characteristic that all large-area electronics share is the low maximum temperature reached during the manufacturing process (typically below 200 °C). This feature is fundamental to integrate circuits on flexible substrates like paper, fabrics and plastic foils without deforming or damaging them, and it is also mandatory to allow innovative processes as printed electronics, which enable at the same time high throughput and low costs.

Semiconductors processable at low temperatures do not have excellent electrical performance (mobilities are about three orders smaller than those in silicon) and cannot easily be doped [1] in order to select the polarity of the charge carriers. For this reason, the most mature processes are usually unipolar, while the manufacturing of different types of TFTs requires the use of different semiconductor materials [2–5]. Also, neither diodes nor linear resistors are normally available. The processing of different semiconducting materials on the same substrate to create

© Springer International Publishing Switzerland 2015   15
D. Raiteri et al., *Circuit Design on Plastic Foils,* Analog Circuits and Signal Processing,
DOI 10.1007/978-3-319-11427-9_3

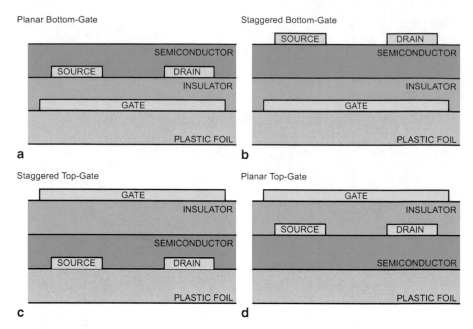

Fig. 3.1 Different thin-film transistor structures

complementary TFT technologies on foil is complex since the first material deposited easily degrades as a consequence of the second deposition. Only recently significant progress has involved complementary technologies. Some of them exploit printed approaches [3] and also provide resistors obtained through the deposition of carbon pastes. Nevertheless these technologies still suffer from the presence of many hard faults and large variability, which hamper the realization of complex circuits; hence, they have not overtaken the most reliable unipolar processes yet. On the contrary, it is presently a matter of strong debate if complementary technologies are really the right choice for the future of circuits manufactured on plastic films.

Unipolar TFTs are typically manufactured through the deposition and patterning of four functional layers: a gate metal layer, a gate insulator layer, a source-drain metal layer and a semiconducting layer. Depending on the order the different layers are processed, four TFT structures can be defined: planar bottom-gate (PBG), staggered bottom-gate (SBG), staggered top-gate (STG) and planar top-gate (PTG) (Fig. 3.1).

Even if staggered structures typically have better contact resistance, since they provide a larger injection surface between source/drain contacts and the semiconductor, it is not possible to simply state which structure is the best. Indeed, on top of carrier injection, several other properties play a role in the overall TFT performance: examples include semiconductor ordering on surfaces with different hydrophobic/hydrophilic behavior or the use of different patterning processes (e.g. printing, photolithography, shadow masking). In general, for one application mobility could be

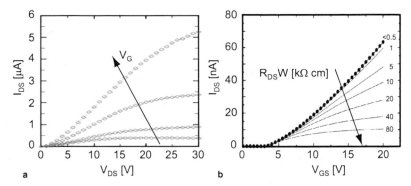

Fig. 3.2 Contact resistance affecting the **a** output (Fig. 6 in [12]) and the **b** transfer characteristic. (Fig. 4 in [13]). (Currents and voltages are shown in line with n-type conventions, even when the device is a p-type.)

the most important figure of merit of a transistor, for another the cut-off frequency or the longevity. Moreover different structures have different costs and suit better different materials. Indeed, the optimal choice should also be tailored to the specific materials and deposition process used for each layer and keeping in mind thermal budget and stack integrity.

For organic TFTs (OTFTs), the PBG structure is very popular, since this structure can achieve shorter channel length and the semiconductor deposition is the last step of the process. This simplifies the manufacturing process and avoids the degradation of the semiconductor due to further processing. First attempts using metal-oxide semiconductors and graphene also exploited the same structure, however we also assisted to the migration to top contact structures. Moreover this structure provides an attractive solution for printed and roll-to-roll technologies, which achieve at the same time the highest throughput and the lowest cost [6, 7].

Unfortunately the short channel length combined with the weak vertical electric field, due to the relatively thick gate insulator (typically above 200 µm), causes a large channel modulation in low-temperature TFTs [8, 9]. Moreover, the small surface available for charge injection between contacts and accumulated channel reflects in a high contact resistance [10, 11]. These two effects considerably degrade the static performance of TFTs and the consequences are clearly visible e.g. in the output and transfer characteristics (Fig. 3.2a, b, respectively).

The interest in TFTs for display backplane applications also encouraged the research on a process providing a third metal layer [14]. Indeed, in some technologies, two additional layers, an insulator and a third metal layer, are deposited atop, taking care to avoid any degradation of the semiconductor and its electrical properties. The most important drive beyond the adoption of this metal layer was in the fact that it enables a very good aperture ratio[1] when fabricating display backplanes, as the pixel driver can be covered with the pixel electrode (Fig. 3.3).

---

[1] The aperture ratio of a pixel is defined as the ratio between the emitting area over the whole surface occupied by the pixel (including wiring, space between adjacent pixels, and pixel driver).

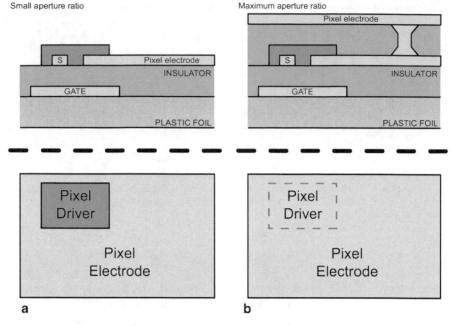

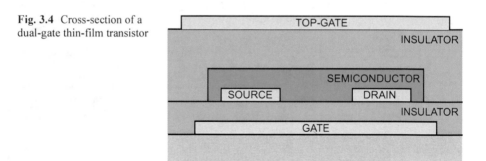

Fig. 3.3 a Low aperture ratio achieved by two metal layers technology and **b** maximum aperture ratio achieved by three metal layers technologies

Fig. 3.4 Cross-section of a dual-gate thin-film transistor

The same three-metal-layers technology can be advantageously used for circuit design, as the third metal layer can be employed as a second gate (Fig. 3.4). This additional TFT control terminal is useful to design innovative circuit topologies that can solve some of the issues mentioned in this work. The second gate indeed influences the electrical properties of the semiconductor underneath and hence a tunable-threshold TFT is obtained. This feature is paramount to achieve at the same time a low-cost process and robust circuit designs, as it will be shown in Chaps. 6–8.

Fig. 3.5 Design flow created
for all the technologies used
for circuit design

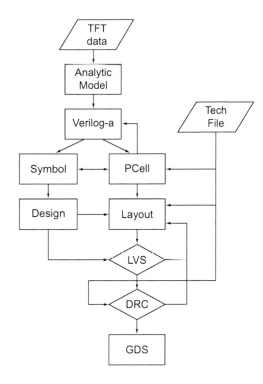

3.2 Design-Environment Setup

The most important application of large-area electronics is in "active matrices" for display backplanes. In this sort of application, TFTs are exclusively employed as switches that provide the right currents or voltages to the pixels in the display. For this reason, a characterization of the TFT transfer and output characteristics with the level of detail that is needed for the design of complex analog and digital circuits is often not provided along with the technology. Beside this, state of the art compact models are often limited to the DC behavior, are not portable among different technologies, and are not openly available in a standard EDA (Electronic Design Automation) environment.

To enable circuit design with TFTs, which is the focus of this work, a physical model for each of the technologies used for our designs was developed in house. To this aim, the state of the art in the field of physical and compact TFT models [15–18] was taken as a reference.

In addition to the lack of a unified portable current model, a very important practical problem is that a proper design kit, implemented in a convenient EDA environment for the simulation, layout and verification of our circuits is almost never available. For this reason, a complete design environment for all the technologies used in this book was set up from scratch in order to provide at least the basic functionalities required during the most important phases of the physical design (Fig. 3.5).

Each TFT technology has been characterized based on the measurements of the transfer and output characteristics of many TFTs with different channel lengths and widths. This is needed to characterize the scaling of the transistor parameters with TFT dimensions and to characterize properties, like carrier injection and short-channel effect, which are strongly dependent on the TFT length. A compact model suitable for each specific technology (e.g. the one described in Chap. 4) is used to extract the parameters, and embedded in a commercial EDA tool by suitably writing it in Verilog-a. The Verilog-a macro is then used within the Spectre simulator. To enable meaningful transient and AC simulations, the main parasitic capacitances were also modeled based on the specific layout geometry of the single device.

The model is then associated to the symbol used in the schematic editor which allows the simulator to fetch the TFT design parameters. These parameters typically define the geometry of the device and can also be read by the PCell (Parameterized Cell). The PCell is used during the layout phase to automatically adjust the dimensions of the different layers accordingly to the design parameter set defined in the simulated schematic and to the layout design rules.

Based on the indications of the technology providers, also the Layout vs Schematic (LVS) tool and the Design Rules Checker (DRC) were configured to verify our designs and achieve a reliable design flow even for complex circuits.

3.3   Circuit Design

In the first large-area applications for flexible electronics, TFTs were mainly used as a switch in active matrices. These switches were used to carry the suitable voltage to pixels in displays or to collect the signal from sensor surfaces [19]. The circuit design was limited to small digital circuits, rectifiers and simple single-stage amplifiers.

The analog and mixed-signal design is a complex design area to tackle with unipolar technologies. Unipolar OTFTs are typically normally-on and this makes impossible to use current mirrors either as an active load (since it should be complementary to the input) or to provide reference bias currents (since the sink TFT does not work in saturation [20], when diode connected). The gain of a single stage voltage amplifier is usually below a few tens [21, 22] and the phase margin decreases rapidly in multistage amplifiers due to the presence of large parasitic capacitances [23]. Of course nested compensation techniques can be used to achieve stability in negative feedback configurations (achieving unity gain bandwidths typically below 2 kHz [24, 25]), but these approaches dramatically increase the circuit complexity with a detrimental effect on the hard/soft faults, and thus yield. For this reason, even discrete-time amplification techniques are difficult to implement. Indeed, a switched-capacitor amplifier requires voltage amplifiers that are internally compensated and have large gain in order to provide a good virtual ground and to switch quickly between the different phases. Moreover, using unipolar normally-on TFTs, charge pumps to turn the devices completely off are required, and transmission gates cannot be implemented. For this reason normally-on TFTs are an extremely bad option to implement switches when all voltage levels must be generated within the circuit itself. In display applications indeed, the drive voltage applied to the switches in the active matrix are applied from outside and exploit e.g. large negative voltages in order to completely switch off the n-type TFTs.

Limitations in the stability of multistage amplifiers are not necessarily an issue in open loop circuits like comparators. Nevertheless, comparators for data converters are difficult to realize due to the large mismatch between devices and to the difficulties in the design of offset-zeroing techniques. A differential pair supplied at 30 V can easily reach an input offset of 1 V [22], hence limiting in principle the linearity of analog to digital converters to less than 5 bit in architectures like flash and pipeline converters which, for this reason, have never been demonstrated.

Other attempts to build analog to digital converters concentrated so far on SAR topologies. As expected, however, the matching of passive devices used to implement the DAC [2, 4] limits the linearity of the source reference signal and hence of the SAR converter. In this case, even exploiting a C-2C approach and external calibration and logic, the resulting converter reaches at the state of the art less than 6 bit linearity [5].

An alternative solution to overcome the issues related to low performance OTAs is to exploit oversampling converters. Unfortunately in this case, the low cut-off frequency of TFTs poses a strong limit to the speed of the comparator, which, at the state of the art, is detrimental for the maximum oversampling factor achievable. For this reason, the $\Delta\Sigma$ modulator reported in [23] achieves an ENOB just a little higher than 4 bits.

In the context of digital design, much more effort has been spent and more relevant achievements have been shown. Digital design with unipolar technologies is however very cumbersome, due to the low intrinsic gain and the availability of only normally-on devices with a single threshold. These limitations result in a poor static behavior of logic gates in terms of gain, symmetry and noise margin. This in turn causes the yield to be too low, which is a concern for single prototypes and much more for mass production of circuits aiming to commercial applications. For this reason, several techniques have been investigated to improve the static characteristics of logic gates manufactured on plastic films, with some benefits and drawbacks.

For instance, in the first plastic RFID demonstrator [26] a diode-load inverter was driven by a level shifter to make the transfer characteristic more symmetrical. Then Pseudo-CMOS logic [27] was applied to many different applications and exploited with different technologies [28–30] but this approach requires three different supply rails. The lowest supply can be used in principle to counteract unavoidable process parameter variations which are the most important cause of soft faults. Nevertheless, being a supply, it is cumbersome to design an automatic correction system on chip.

The most impressive results were anyway achieved taking advantage of a double-gate technology. The dual-gate enhanced logic style [31] was indeed used to fabricate the first organic microprocessor [32], counting more than 3000 transistors on the same foil. This approach exploits the second gate to control the threshold voltage of the pull-up network. Unfortunately the second gate has a weak influence on that parameter. Therefore, voltages even higher than three times the supply (typically 20 V) are required.

In the next chapter, this very same double gate technology will be described more in detail, as it will be used extensively for analog, digital and mixed-signal designs in this work. It is firm belief of the authors that a low-defect unipolar technology providing dual-gate feature is the only way to achieve low cost, high density (compared to other large-area electronics solutions) and high yield circuits, all factors needed to enable the adoption of such technology in real life applications.

References

1. B. Lüssem, M. Riede, K. Leo, Doping of organic semiconductors. Phys. Status Solidi A **210**(1), 9–43 (2013)
2. T. Zaki et al., A 3.3 V 6-bit 100 kS/s current-steering digital-to-analog converter using organic p-type thin-film transistors on glass. IEEE J. Solid-State Circuits. **47**(1), 292–300 (2012)
3. S. Jacob et al., High performance printed N and P-type OTFTs for complementary circuits on plastic substrate. *IEEE European Solid-State Device Research Conference,* pp. 173–176, 2012
4. W. Xiong, Y. Guo, U. Zschieschang, H. Klauk, B. Murmann, A 3-V, 6-Bit C-2C digital-to-analog converter using complementary organic thin-film transistors on glass. IEEE J. Solid-State Circuits. **45**(7), 1380–1388 (2010)
5. W. Xiong, U. Zschieschang, H. Klauk, B. Murmann, A 3 V 6b successive-approximation ADC using complementary organic thin-film transistors on glass. *IEEE International Solid-State Circuits Conference*, pp. 47–49, 2010
6. M.J. Yu et al., Amorphous InGaZnO thin-film transistors compatible with roll-to-roll fabrication at room temperature. IEEE Electron Device Lett. **33**(1), 47–49 (2012)
7. H. Zhang, M.D. Poliks, B. Sammakia, A roll-to-roll photolithography process for establishing accurate multilayer registration on large area flexible films. J. Display Technol. **6**(11), 571–578 (2010)
8. N.H. Touidjen, F. Mansour, Modeling kink effect in the poly-Si TFTs under charge sheet approach. *International Conference on Advances in Computational Tools for Engineering Applications*, pp. 407–410, 2009
9. L. Mariucci et al., Polysilicon TFT structures for kink-effect suppression. IEEE Trans. Electron Devices. **51**(7), 1135–1142 (2004)
10. G.B. Blanchet, C.R. Fincher, M. Lefenfeld, J.A. Rogers, Contact resistance in organic thin film transistors. Appl. Phys. Lett. **84**(2), 296–298 (2004)
11. H. Klauk et al., Contact resistance in organic thin film transistors. Solid State Electron. **47**(2), 297–301 (2003)
12. C.H. Kim, Y. Bonnassieux, G. Horowitz, Charge distribution and contact resistance model for coplanar organic field-effect transistors. IEEE Trans. Electron Devices. **60**(1), 280–287 (2013)
13. P. Servati, D. Striakhilev, A. Nathan, Above-threshold parameter extraction and modeling for amorphous silicon thin-film transistors. IEEE Trans. Electron Devices. **50**(11), 2227–2235 (2003)
14. G.H. Gelinck, E.v. Veenendaal, R. Coehoorn, Dual-gate organic thin-film transistors. Appl. Phys. Lett. **87**(7), 073508 (2005)
15. E. Calvetti, L. Colalongo, Z.M.K. Vajna, Organic thin film transistors: an analytical model for circuit simulation. *IEEE Region 10th Conference*, pp. 306–309, 2004

16. E. Calvetti, A. Savio, Z.M. Kovács-Vajna, L. Colalongo, Analytical model for organic thin-film transistors operating in the subthreshold region. Appl. Phys. Lett. **87**(22), 223506 (2005)
17. M. Estrada et al., Mobility model for compact device modeling of OTFTs made with different materials. Solid State Electron. **52**(5), 787–794 (2008)
18. O. Marinov, M.J. Deen, U. Zschieschang, H. Klauk, Organic thin-film transistors: part I—compact DC modeling. IEEE Transact. Electron Devices. **56**(12), 2952–2961 (2009)
19. H.E.A. Huitema et al., Plastic transistors in active-matrix displays. Nature. **414**(6864), 599–599 (2001)
20. D. Raiteri, F. Torricelli, E. Cantatore, A.H.M.V. Roermund, A tunable transconductor for analog amplification and filtering based on double-gate organic TFTs. *IEEE European Solid-State Circuits Conference*, pp. 415–418, 2011
21. N. Gay, W.J. Fischer, OFET-based analog circuits for microsystems and RFID-sensor transponders. *International Conference on Polymers and Adhesives in Microelectronics and Photonics*, pp. 143–148, 2007
22. M.G. Kane et al., Analog and digital circuits using organic thin-film transistors on polyester substrates. IEEE Electron Device Lett. **21**(11), 534–536 (2000)
23. H. Marien, M.S.J. Steyaert, E. van Veenendaal, P. Heremans, A fully integrated ΔΣ ADC in organic thin-film transistor technology on flexible plastic foil. IEEE J. Solid-State Circuits. **46**(1), 276–284 (2011)
24. H. Marien, M. Steyaert, N.V. Aerle, P. Heremans, A mixed-signal organic 1 kHz comparator with low VT sensitivity on flexible plastic substrate. *IEEE European Solid-State Circuit Conference*, pp. 120–123, 2009
25. G. Maiellaro et al., High-gain operational transconductance amplifiers in a printed complementary organic TFT technology on flexible foil. IEEE Transact. Circuits Syst. Regul. Pap. **60**(12), 3117–3125 (2013)
26. E. Cantatore et al., A 13.56 MHz RFID system based on organic transponders. IEEE J. Solid-State Circuits. **42**(1), 82–92 (2007)
27. T.C. Huang et al., Pseudo-CMOS: a design style for low-cost and robust flexible electronics. IEEE Transact. Electron Devices **58**(1), 141–150 (2011)
28. K. Ishida et al., Insole pedometer with piezoelectric energy harvester and 2 V organic circuits. IEEE J. Solid-State Circuits **48**(1), 255–264 (2013)
29. K. Ishida et al., A 100-V AC energy meter integrating 20 V organic CMOS digital and analog circuits with a floating gate for process variation compensation and a 100 V organic pMOS rectifier. IEEE J. Solid-State Circuits. **47**(1), 301–309 (2012)
30. T. Yokota et al., Sheet-type flexible organic active matrix amplifier system using pseudo-CMOS circuits with floating-gate structure. IEEE Trans. Electron Devices. **59**(12), 3434–3441 (2012)
31. K. Myny et al., Unipolar organic transistor circuits made robust by Dual-Gate technology. IEEE J Solid-State Circuits. **46**(11), 1223–1230 (2011)
32. K. Myny et al., An 8-bit, 40-instructions-per-second organic microprocessor on plastic foil. IEEE J. Solid-State Circuits. **47**(1), 284–291 (2012)

Chapter 4
Device Modeling and Characterization

Any reliable circuit design is based on compact models able to describe accurately the behavior of the TFTs in a given technology. With our modeling, accurate simulation of analog circuits in a CAD platform is enabled and deeper insight in the underlying transport mechanisms is achieved. Thus, a physical model was developed in house that guarantees the continuity and the derivability among the different working regions, ensuring as well the symmetry between source and drain in the channel current. In this chapter, the implemented model will be explained first, and characterized afterwards. Parts of this chapter have been published in [1, 2].

4.1 Transistor Model

In the previous chapters, it has often been highlighted that semiconductors used for low-temperature electronics have worse electrical performance compared to silicon technologies. The lower mobility and uniformity are intrinsic consequences of the non mono-crystalline nature of the semiconductor, which is a consequence of the low-temperature deposition methods (as solution processing and RF sputtering) that are suitable to flexible substrates. For this reason, higher mobilities have been measured in TFTs processed at low temperatures, where a self-assembled monolayer (SAM) was deposited before the semiconductor [3]. A SAM can be indeed used to quickly produce a structured substrate and favor ordering of the semiconducting organic molecules [4].

In mono-crystalline semiconductors, atoms in the regular crystal structure interact creating allowed and forbidden energy levels for their electrons, i.e. bands and gaps respectively. Due to the crystalline ordering, these levels are shared over the whole crystal and for all charge carriers, hence they are referred to as *extended* states. With increasing material disorder, the well-known concept of bands (as the valence and the conduction ones) separated by an energy gap needs to be reconsidered, since different phenomena have to be taken into account. For instance, opposite to crystal lattice are polymer materials, where the molecular disorder completely hampers the

© Springer International Publishing Switzerland 2015
D. Raiteri et al., *Circuit Design on Plastic Foils,* Analog Circuits and Signal Processing,
DOI 10.1007/978-3-319-11427-9_4

Fig. 4.1 Multiple trapping
and release transport in
metal-oxide and amorphous
semiconductors. © 2011
IEEE. Reprinted, with
permission, from F. Torricelli
et al. [5]

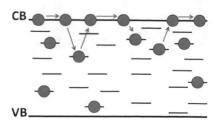

generation of globally shared energy levels, and charge carriers can only occupy
physically *localized* states. In the light of this consideration, also the charge transport has to be supported by different theories.

In the case of amorphous metal-oxides materials, the semiconductor is typically
described as a system consisting of valence and conduction bands; however localized energy states in the energy gap can trap the charge carriers. If the majority
of the charge carriers remains trapped in these localized energy states, the charge
transport in the semiconductor can be modeled by means of the Multiple Trapping
and Release (MTR) theory [5]. According to the MTR, a trapped carrier can be
activated thermally (Fig. 4.1), and move in the conduction band (CB) until it falls
in a trap again.

In most organic semiconductors, typically bands are not associated to the system.
In this scenario, all the free carriers are trapped in localized states and the charge
transport can take place by "charge hopping" between different localized states.
This mechanism is referred to as Variable Range Hopping (VRH) [7]. According to
VRH, the probability of a hop event depends on the phonon frequency, on the difference in energy between the initial and final state (which can either favor or disfavor
the hop), and on the spatial separation between the states (Fig. 4.2). Other factors
also play a role in the hopping probability: the Density of States (DOS) at different
energy levels, the free carrier wavefunction, etc. All these factors need to be taken
into account when evaluating the conductivity of the semiconductor and writing the
expression of the static currents in organic TFTs.

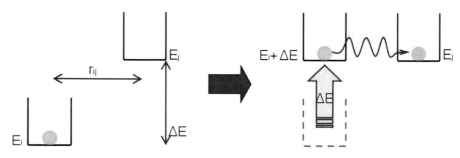

Fig. 4.2 Variable range hopping transport in organic semiconductors [6]

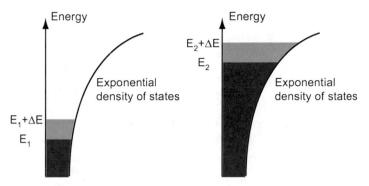

Fig. 4.3 Localized states with respect to energy level of the semiconductor

Although the DOS for disordered materials governed by VRH is typically assumed to be Gaussian [8], most models adopt an exponential relation between the number of states and the given energy

$$g(E) = N_t \, exp\left(\frac{E}{K_B T_0}\right)\theta(-E) \tag{4.1}$$

this is a reasonable approximation since the energies involved typically concern the tails of the Gaussian bell, which are very well approximated by an exponential function. In this expression, N_t is the total number of localized/trap states, K_B the Boltzmann constant, T_0 the characteristic temperature (related to the width of the exponential distribution, which accounts for the level of disorder) and θ is the Heaviside unit step function ($\theta(x) = 1$ if $x > 0$, otherwise $\theta = 0$).

This DOS shape has very important consequences in the physical understanding of the current law that will be eventually derived. Let us consider, indeed, two different bias conditions (Fig. 4.3) such that all localized states are filled up till the energy level E_1 in the first case and up to E_2 in the second, with $E_1 < E_2$. In the first case, the number of carriers that can be thermally activated is much lower than in the second case, and, most important, increasing the base energy level by the same energy step ΔE, an exponentially larger number of localized states can be reached. Due to this phenomenon, the hop probability rises dramatically in the case where the charge carriers have higher energy. Hence, the mobility is not constant, but increases with the energy level of the occupied states in the semiconductor. In a field effect transistor, this means that the mobility increases with the voltage applied to the gate.

In order to better understand the mathematical expression of the current, besides MTR and VRH, it is useful to illustrate the most important difference between silicon and unipolar thin-film transistor technologies on foil. TFTs on foil typically work in accumulation and not in inversion. Therefore, even if the semiconductor layer is intrinsic, it causes a major source of leakage current between source and drain, with a detrimental effect on the on–off current ratio. On the other hand, accumulation TFTs are typically normally-on, which is a very important feature

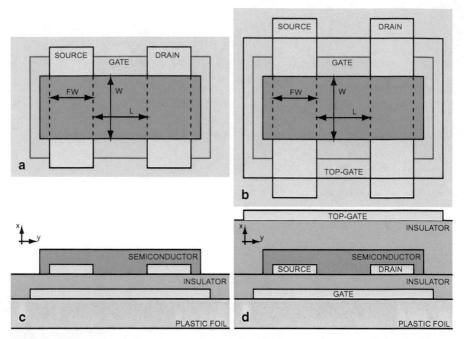

Fig. 4.4 a, b Top-view and **c, d** cross-section of a thin-film transistor manufactured with **a–c** single and **b–d** double-gate technologies

to enable reliable circuit design with unipolar technologies, as it will be shown in Chaps. 6–8. The recurring issue of organic transistors stability in air is also due to this aspect of large-area technologies. Indeed water molecules in the air, are inclined to interact with the semiconductor, and behave like dopants both worsening mobility and threshold voltage (i.e. the on current), and increasing the leakage in the substrate (i.e. the off current). Moreover, since leakage can take place, in a real circuit, also between nearby transistors, the semiconductor is typically patterned and sized within the gate (Fig. 4.4a) and, when available, within the top-gate (Fig. 4.4b).

In order to find an expression for the TFT channel current, let us consider first the most widely used TFT structure used for large-area electronics, namely the planar bottom-gate one (Fig. 4.4c). The case of TFTs with double gate (Fig. 4.4d) will be investigated later. Even if the technology here characterized is p-type-only, for the sake of simplicity the model will be written assuming n-type TFTs. According to the approach in [6], using the coordinate system shown in Fig. 4.4, the current above threshold can then be derived considering accumulated carriers at the interface between insulator and semiconductor ($x=0$), and a current flowing in the y direction from the source ($y=0$, $V_{ch}=V_S$) to drain edge ($y=L$, $V_{ch}=V_D$). Applying the drift-diffusion relationship:

$$J_n(x,y) = qn\mu_n \boldsymbol{E} + qD_n \nabla n \simeq \sigma(x,y)\frac{dV_{ch}}{dy} \qquad (4.2)$$

where σ is the conductivity of the semiconductor, which takes into account the specific DOS and transport mechanism, and V_{ch} is the channel potential. The diffusion term will be neglected here, as a biasing above the sub-threshold region will be assumed, where the main current contribution is due to drift.

The above threshold channel current can be modeled following the standard approach used also in silicon technologies. Equation (4.2) needs to be integrated in both the x and y directions to obtain the drain-source current

$$I_{DS} = \frac{W}{L} \int_{V_S}^{V_D} \int_0^{t_p} \sigma(x, y) \, dx \, dV_{ch}. \tag{4.3}$$

Here, W and L are respectively the channel width and the channel length of the device and t_p is the thickness of the semiconductor. After the proper math [9] on the expression of the conductivity, it is possible to write the drain-source current as a function of the charge accumulated in the channel at the source and at the drain contacts (Q_S and Q_D):

$$I_{DS} \simeq \frac{W}{L} \frac{\zeta}{C_i} \frac{T}{T_0} \left[Q_S^{\frac{2T_0}{T}} - Q_D^{\frac{2T_0}{T}} \right] \tag{4.4}$$

where ζ includes all the electrical parameters characterizing the transport, C_i is the insulator capacitance per unit area between the semiconductor and the gate, T is the working temperature of the transistor, and T_0 is a characteristic temperature related to the disorder of the system, i.e. defining the exponential DOS.

In order to express the static current as a function of the voltage applied to the terminals, the charge accumulated at the insulator-semiconductor interface can be written as:

$$Q_X = C_i(V_G - V_{FB} - V_X) \tag{4.5}$$

where V_G and V_X are the voltages applied respectively to the gate and to the X terminal (source or drain). The voltage V_{FB} is an equivalent flat-band voltage that accounts for the charge neutrality in the metal-insulator organic-semiconductor structure. The model for the drain source current is thus:

$$I_{DS} \simeq \frac{1}{2} \frac{W}{L} \frac{2T}{T_0} \zeta C_i^{\frac{2T_0}{T}-1} \left[(V_G - V_{FB} - V_S)^{\frac{2T_0}{T}} - (V_G - V_{FB} - V_D)^{\frac{2T_0}{T}} \right]. \tag{4.6}$$

It is worth noting that the ratio $2T_0/T$ is typically larger than 2. The cause of this exponent in the dependence of the current from the gate voltage is in the variation of the mobility with the gate voltage, which has been discussed above.

This formulation of the channel current holds both for triode and for saturation regions, if the factor $V_G - V_{FB} - V_D$ is assumed equal to zero when it would become negative, but it does not take into consideration the channel length modulation effect. For this purpose, an additional scaling factor K_{sat} can be included in the drain current expression, paying attention to preserve the continuity of the characteristic and its derivatives between the linear and the saturation regions. Also in this case, the procedure to evaluate this additional term is the same as for silicon technologies and it can be expressed as:

$$K_{sat} = 1 + \frac{V_{DS}}{E_p L} \tag{4.7}$$

where the equivalent Early voltage V_p is expressed as the product of the pinch-off field E_p and the channel length L, to enable scaling with the dimensions of the device.

The last two biasing conditions that still need to be taken into account are the sub-threshold region and the finite off-current due to the resistance of the bulk (remind that, as explained before, these devices work in accumulation). The latter can be simply modeled with a shunt resistor between source and drain that scales with the geometrical dimensions of the transistor:

$$I_{off} = \frac{W}{L} \frac{V_{DS}}{R'_{sub}} . \tag{4.8}$$

The sub-threshold behavior of these devices is exponential: some attribute the exponential behavior to the diffusion term, others explain it with the dependence of the charge carrier injection probability with the gate voltage. Indeed, the gate voltage strongly affects the energy of the semiconductor in the sub-threshold region and consequently it modifies the barrier height at the metal-semiconductor interface. In any case, the exponential current in the subthreshold region can be empirically modeled writing the overdrive voltages on the source and drain sides ($V_{OD,X} = V_G - V_X - V_{FB}$) as:

$$V_{OD,X}(V_X) = V_{SS} ln \left[1 + exp \left(\frac{V_G - V_{FB} - V_X}{V_{SS}} \right) \right]. \tag{4.9}$$

This mathematical approach allows to fit the sub-threshold slope to the characteristic of the device using the parameter V_{SS}, and preserves the continuity and derivability of the characteristic above threshold.

In the presence of a second gate, all the above considerations still hold. On top of that, the voltage applied to the top-gate also has an influence. If an accumulation layer is not created between top insulator and the semiconductor, the electric field generated by the top-gate interacts with the one generated by the bottom gate, shifting the onset of accumulation in the channel. Due to the different thickness of the top-gate and bottom gate insulator layers, the top-gate voltage needs to be properly weighted [10]. In presence of the second gate, thus, the V_{FB} used in Eqs. (4.5), (4.6), and (4.9) should be substituted by:

$$V_T = V_{FB} - \eta (V_{TG} - V_S) \tag{4.10}$$

where the scale factor η is the ratio between the thickness of the bottom-gate and top-gate insulator [11].

4.2   Measurements and Characterization

Once the physical model of the current that applies to our technology is defined, the proper value of each parameter needs to be determined based on the actual device measurements. For this purpose, a comprehensive set of devices was designed with different channel lengths and widths. The channel length of the devices was scaled from 5 (the minimum feature size) to 100 μm in order to investigate contact effects, channel properties and device scalability. On the other hand, in order to measure suitable currents, the channel width was scaled accordingly.

Due to the presence of four terminals, three types of characteristics were measured:

- Bottom-gate transfer characteristic (TC)
- Top-gate transfer characteristic (TCTG)
- Output characteristic (OC)

For each case, the independent variable was swept forward and backward in order to identify possible hysteresis and other stress effects, while many measurements have been performed for different biases applied to the remaining three terminals. An example of the three characteristics is reported in Fig. 4.5.

Despite the variety of phenomena occurring in the OTFT, as discussed in the previous section, the current model can be expressed by means of a compact expression including only seven parameters.

$$I_{DS} = \frac{W}{L}\beta\left[V_{OD,S}{}^{\gamma} - V_{OD,D}{}^{\gamma}\right]\left(1 + \frac{V_{DS}}{E_P L}\right) + \frac{W}{L}\frac{V_{DS}}{R'_{sub}} \qquad (4.11)$$

$$V_{OD,X} = V_{SS} ln\left[1 + exp\left(\frac{V_G - V_X - V_{FB} + \eta(V_{TG} - V_S)}{V_{SS}}\right)\right] \qquad (4.12)$$

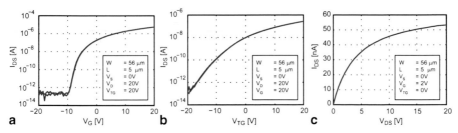

Fig. 4.5 a Bottom-gate and **b** top-gate transfer characteristics, and **c** output characteristic of a double gate OTFT with W = 56 μm and L = 5 μm

Table 4.1 Model parameters for the double-gate p-type organic technology

	Symbol	Value	Unit
Current prefactor	β	3×10^{-10}	$[A/V^\gamma]$
Traps coefficient	γ	2.37	[K/K]
Sub-threshold slope	V_{SS}	1	[V]
Flat-band voltage	V_{FB}	1.2	[V]
Bulk resistance	R'_{SUB}	$1 \times 10^{+14}$	$[\Omega]$
Pinch-off field	E_p	$1.5 \times 10^{+5}$	[V/cm]
Top-gate coupling	η	0.25	$[\mu m/\mu m]$

where the overdrive voltage has been written separately for the sake of readability. All needed parameters are listed in Table 4.1, together with their typical value for the p-type-only organic technology used in this work and their units.

In order to tailor the model to suit our specific technology, a comprehensive set of routines, described using Matlab, was applied to all measured data. The first step to quickly analyze thousands of measured characteristics was the creation of a uniform scheme for data storage (Fig. 4.6). For each device, a structured variable was created containing the relevant information about the device (foil, technology, tapeout date, et cetera), all the necessary parameters, and all the measured data (organized by characteristic type). In the multidimensional array, each plane contains the current and voltage values measured or applied to each terminal during the measurement of a single characteristic. In this plane, each row corresponds to a point of the characteristic (red box in Fig. 4.6), hence for each row Kirchhoff's laws can be verified and used to check the data consistency. Different planes are data collections that have been retrieved for instance at different times or different bias. Accordingly to the characteristic type each plane is then stored in the right sub-variable (OC, TC or TCTG).

Based on this data structure, a specific value for each parameter listed in Table 4.1 was identified for each measured device. The output characteristic was used to detect the pinch-off electric field E_p, the top-gate transfer characteristic to estimate the top-gate coupling factor η and the bottom-gate transfer characteristic for all the other parameters.

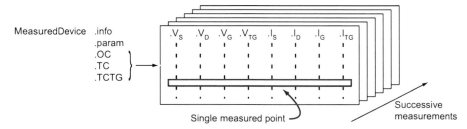

Fig. 4.6 Structured variable used in Matlab for quick semi-automated characterization

The extraction routine also allows to store in the proper location the values of the parameters that fit the model to each device. Due to the variability of the technology, every device has a different set of parameters. Hence, in order to choose the value which is most representative for the technology, the median value of each parameter was picked. The median provides the best statistical estimate of the mean value when the central limit theorem cannot be applied due to a limited data set, since it effectively excludes the outliers.

4.2.1   Parameters Extraction

In this section, the extraction of the parameters for a typical device will be discussed. For the sake of synthesis and simplicity, the case of transfer characteristics measured in the saturation region will be detailed, however similar considerations hold if the device under test works in its linear region. The semi-automatic routine can detect the correct algorithms to be applied based on the fetched bias of the device and on the extracted threshold voltage.

The only parameter that can be extrapolated from the output characteristics of the device under test is the Early voltage V_p. For a low bias of V_G and of V_{TG}, and around the highest drain voltages applied to measure the OC, the TFT surely works in saturation. In that region (see Fig. 4.7), the OC can be extrapolated with a straight line crossing the horizontal axis in $V_{DS} = -V_p$. In order to extract a parameter value independent of the channel length, the equivalent pinch-off electric field E_p can be obtained simply dividing V_p by the channel length of the measured device.

The first parameters that can be extracted from the transfer characteristic of a device are the traps coefficient γ and the threshold voltage V_T. For this purpose, can be used the function $w = w(V_{GS})$ defined as[12]:

$$w = \frac{\int I_{DS} dV_{GS}}{I_{DS}}. \tag{4.13}$$

Fig. 4.7 The *blue line* shows the measured OC used for the evaluation of the pinch-off field E_p. The *red line* fits the saturation region of the characteristic and allows the estimation of the Early voltage V_p. *Dashed lines* show the interpolation range

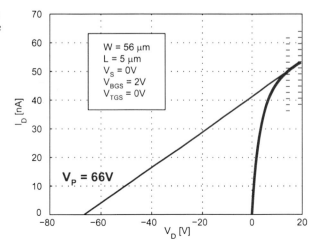

Fig. 4.8 Traps coefficient evaluation. The *blue line* represents the function $w = w(V_{GS})$ (Eq. (4.13)) based on the measured TC in Fig. 4.5a. The *red line* is the linear fit in the saturation region, between 0 and 15 V. *Dashed lines* show the interpolation range

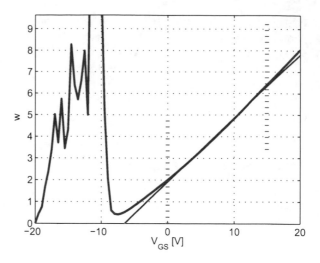

Indeed, when the TFT works in saturation, using Eqs. (4.11) and (4.13) w can be approximated by:

$$w_{sat} = \frac{V_{GS}}{\gamma + 1} + \frac{V_T}{\gamma + 1}. \tag{4.14}$$

In Fig. 4.8 the evaluation of w based on a measured TC is shown. In the sweep region between $V_{GS} = 0$ V and $V_{GS} = 15$ V the OTFT works in saturation since $V_{DS} - V_{GS} > V_T$ and $w \simeq w_{sat}$. Approximating $w = w(V_{GS})$ with a straight line $y = mV_{GS} + q$ (red line in Fig. 4.8), can be immediately derived

$$\gamma = \frac{1}{m} - 1. \tag{4.15}$$

$$V_T = -\frac{q}{m}. \tag{4.16}$$

It is worth noting that this threshold voltage is not one of the seven parameters, but it is a function of V_{FB} and η, whose values will be evaluated later in this section.

In a way similar to the evaluation of γ, it is possible to estimate the prefactor β. In this case can be used the function:

$$z = \frac{\int I_{DS}^2 dV_{GS}}{I_{DS}}. \tag{4.17}$$

Over the same gate voltage range where the traps coefficient has been extracted, we can also identify a $z_{sat} = z_{sat}(V_{GS})$ approximation of $z = z(V_{GS})$:

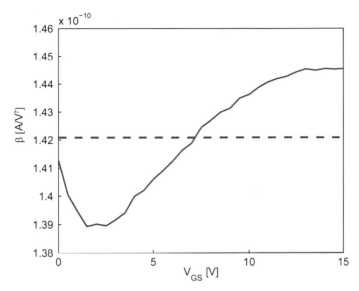

Fig. 4.9 Evaluation of the prefactor β. The *continuous line* is obtained through the manipulation of the z function, while the *dashed line* is its average within the range of interest

$$z_{sat} = \frac{W}{L}\frac{\beta}{\gamma+1}(V_{GS} - V_T)^{\gamma+1}. \tag{4.18}$$

Therefore, dividing the z_{sat} by $\frac{W}{L}\frac{1}{\gamma+1}(V_{GS} - V_T)^{\gamma+1}$, the value of the prefactor β is obtained. The value of β as a function of V_{GS} is shown in Fig. 4.9 along with its average value (dashed line).

For the evaluation of the sub-threshold slope V_{SS}, the first aspect to highlight is that the slope of the logarithm of the sub-threshold current (see TC of Fig. 4.10) is, in the case of saturation regime, proportional to $1/V_{SS}$. Indeed in saturation, the Eqs. (4.11) and (4.12) of the drain-source current can be simplified to:

$$I_{DS,SS} \simeq kV_{SS}^{\gamma}ln\left[1+exp\left(\frac{V_{GS}-V_T}{V_{SS}}\right)\right]^{\gamma} \tag{4.19}$$

where k includes the geometry of the device W/L, the prefactor β and the channel length modulation factor (V_{DS} is constant during the measurement of the TC). Moreover, the argument of the exponential is much smaller than zero, hence the logarithmic expression can be replaced with the exponential only ($ln(1+x) \simeq x$ when $x \simeq 0$) resulting in:

$$I_{DS,SS} \simeq kV_{SS}^{\gamma}exp\left(\gamma\frac{V_{GS}-V_T}{V_{SS}}\right). \tag{4.20}$$

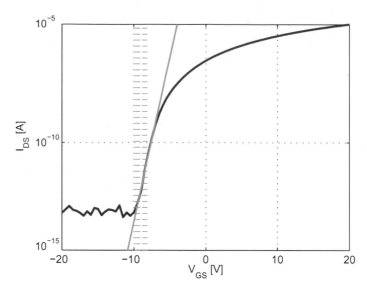

Fig. 4.10 Evaluation of the sub-threshold slope V_{SS}

Applying the base 10 logarithm to both sides, a linear function of V_{GS} in the semi-logarithmic plane $Log(I) - V_{GS}$ is obtained:

$$Log(I_{DS,SS}) \simeq \gamma Log(kV_{SS}) + \frac{\gamma Log(e)}{V_{SS}}(V_{GS} - V_T). \tag{4.21}$$

The green line of Fig. 4.10 ($y = mV_{GS} + q$) approximates the characteristic exactly in the region of interest, hence the sub-threshold slope V_{SS} can be expressed through its angular coefficient m and reads:

$$V_{SS} = \frac{\gamma Log(e)}{m}. \tag{4.22}$$

The top-gate coupling parameter models the effect of the voltage applied to the top-gate on the drain-source current. Therefore, this parameter was evaluated based on the top-gate transfer characteristic. In this case, for each measured point within the top-gate sweep, the value of η is found such that the modeled current equals the measured one. However, since the extracted value of the flat-band voltage V_{FB} is still missing, it was expressed as a function of V_T and η:

$$V_{FB} = V_T + \eta(V_{TG}^* - V_S) \tag{4.23}$$

where V_{TG}^* is the voltage applied to the top-gate when the TC used to evaluate γ and V_T was measured. Since the TCTG is measured in saturation, the modeled

Fig. 4.11 Evaluation of the coupling factor η. The *thick line* represents H evaluated from measured data, while the *thin line* represents a credible extension of H for an input range larger than V_{TG}^*

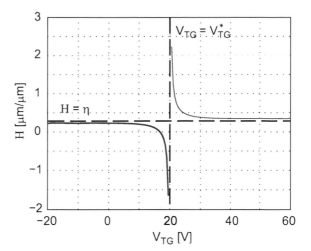

drain-source current, i.e. Eqs. (4.11) and (4.12), to be compared with the measured TCTG, can be rewritten as:

$$I_{DS} = k \left[V_G - V_S - V_T + \eta(V_{TG} - V_{TG}^*) \right]^{\gamma} \tag{4.24}$$

where k is the same prefactor as used in Eq. (4.19). Aware of Eq. (4.23), η can be evaluated using a functional H defined as:

$$H = \frac{\sqrt[\gamma]{I_{TCTG}/k} - V_{GS} + V_T}{\left(V_{TG} - V_{TG}^* \right)} \tag{4.25}$$

where I_{TCTG} represents the measured data. Figure 4.11 plots H obtained from the TCTG data shown in Fig. 4.5b and for a threshold voltage V_T extracted applying $V_{TG}^* = 20$ V. Equation (4.25) draws a hyperbola in the H-V_{TG} plane with asymptotes in $V_{TG} = V_{TG}^*$ and H=η (i.e. value to be used in the final model). Since the measured data in Fig. 4.5b have been obtained for a top-gate voltage range lower than the vertical asymptote, $\eta = H(V_{TG} = -20$ V) was chosen. In the case of an input voltage range sweeping across the vertical asymptote, η can be computed as the average between the extreme values $\eta_1 = H(V_{TG, min})$ and $\eta_2 = H(V_{TG, max})$.

The last parameter to be extracted is the bulk resistance, R'_{sub}. Its value can be estimated by the off-current I_{off} measured in a TC (Fig. 4.12). The only remark about this parameter is that many devices should be measured in order to be sure that the lowest value of the current is indeed the off-current of the transistor, scaling accordingly to W and L, and not the noise related to the measurement setup. Referring to Fig. 4.12 and in line with Eq. (4.8), R'_{sub} is given by:

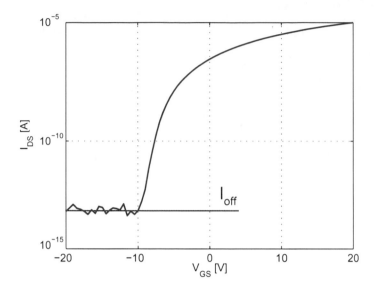

Fig. 4.12 Evaluation of the bulk resistance R'_{sub}

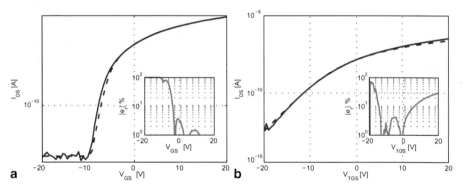

Fig. 4.13 Comparison of the characterized model (*red dashed line*) with the measured **a** TC and **b** TCTG (*blue line*). The insets show the percentage error

$$R'_{sub} = \frac{L}{W} \frac{V^*_{DS}}{I_{off}}$$ (4.26)

where V^*_{DS} is the voltage applied between drain and source during the measurement of the TC.

After the extraction of all parameters, the agreement between the measured data and the model can be evaluated. Figure 4.13 shows the transfer characteristics sweeping both bottom-gate and top-gate (blue lines) together with the modeled characteristic (red dashes). In the inset, the error percentage is shown on a logarithmic scale.

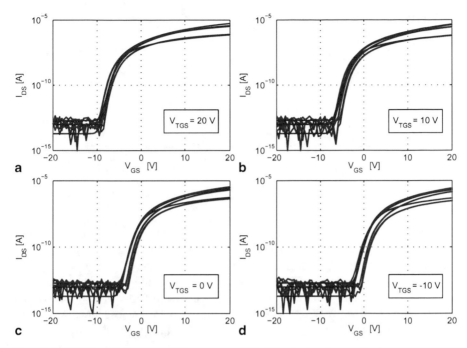

Fig. 4.14 Model validation over all the measured TC of the device characterized

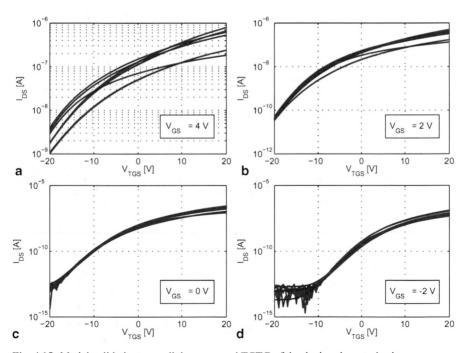

Fig. 4.15 Model validation over all the measured TCTG of the device characterized

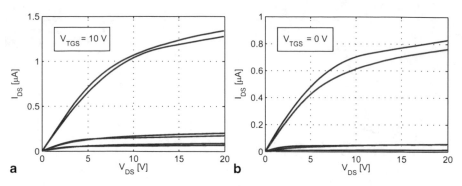

Fig. 4.16 Model validation over all the measured OC of the device characterized

The model was then validated for all the measurements performed on the same device. In Fig. 4.14 each panel shows, together with the top-gate voltage, the TCs measured for three different drain-source voltages ($V_{DS} = 2$, 10 and 20 V). Figure 4.15 shows different TCTGs obtained with the same drain-source voltages as before and for four different gate-source voltages (reported in the inset). The OCs measured for $V_{GS} = 0$, 2, 10 V are shown in Fig. 4.16. In these three Figs. 4.14, 4.15, 4.16, the measured data are represented with a blue line, while the model is plotted in red.

4.2.2   Dynamic Behavior

The characterization of the technology so far illustrated only takes into account the static properties of the transistor behavior. In order to perform transient simulations however, overlap capacitances between gate, source and drain (Fig. 4.17) have also been introduced in the model. Both gate-source and gate-drain capacitors consist of

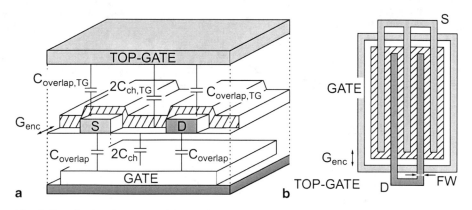

Fig. 4.17 a Stack and **b** layout of the implemented double-gate TFTs

a constant component $C_{overlap}$ due to the overlap between gate and contact fingers (much larger than in self-aligned silicon technologies), and of a variable component depending on the channel accumulation state.

These parasitic capacitors influence the dynamic behavior of the TFTs. Also for TFTs manufactured at low temperature, the cut-off frequency is proportional to the ratio between transconductance g_m and input capacitance C_i [13]:

$$f_i = \frac{g_m}{2\pi C_i}.$$
(4.27)

However, contrary to self-aligned technologies, overlap capacitances are not negligible and they should also be taken into account when evaluating the maximum cut-off frequency, especially for minimum channel length devices. For this reason, a scale factor, depending on the number of sub-channels SC, needs to be included to take into account the source and drain overlap capacitances. Since the minimum finger width FW and the minimum channel length L typically have same dimensions (FW = L_{min} = 5 μm for our lithography process), the maximum cut-off frequency can be expressed as:

$$f_{i,max} = \frac{g_m}{2\pi C_i} \simeq \frac{SC}{2SC+1} \frac{\mu(V_G - V_T)}{2\pi L_{min}^2}$$
(4.28)

where the scaling factor depending on SC ranges between 1/3 and 1/2. The relatively thick gate dielectric used in TFTs on foil (from 100 nm to 1.5 μm typically) requires overdrive voltages in the order of 10 V to obtain sufficient currents. Therefore, large-area TFT cut-off frequencies are limited at the state of the art to about 200 kHz (the most advanced silicon MOSFETs have cut-off frequencies in excess of hundreds of GHz).

The final capacitance model considers in detail the device geometry specified in the parameterized cell (PCell). For instance, the PCell developed for our CAD system only accepts an even number of sub-channels SC (see Fig. 4.17b), leading thus to a smaller gate-drain capacitance C_{GD} compared to the gate-source C_{GS} one:

$$C_{GS} = C_{GD}\left(1 + \frac{2}{SC}\right).$$
(4.29)

In a double-gate TFT, the parasitic capacitances between top-gate and channel, source and drain have, in first approximation, the same geometry as the ones associated to the gate terminal. However, the different thickness of the insulator requires an additional scale factor which is equal to the top-gate coupling factor η. Therefore, the top-gate capacitances result:

$$C_{TGX} = \eta C_{GX}$$
(4.30)

where the subscript X stands for source (S), drain (D) or channel (ch). All the required parasitic capacitances are taken into account by the Verilog-a model, simply including a suitable capacitance between each terminal pair.

The final model was implemented in the environment setup and used for static and transient simulations, which are fundamental to the design of the circuits shown in the next chapters. The statistics concerning matching and variability are not available yet from the manufacturers, and these aspects of the modeling fall out the scope of this work. However in the future, improved versions of the TFT model should integrate the characterization of process variations. To make Monte Carlo simulations possible, a Spectre model including the Verilog-a one can be used. Spectre is able to generate automatically, for each transistor in the schematic, a unique parameter set in line with the measured average and the standard deviation of each parameter in the Verilog-a model.

References

1. D. Raiteri, F. Torricelli, E. Cantatore, A.H.M.V. Roermund, A tunable transconductor for analog amplification and filtering based on double-gate organic TFTs. *IEEE European Solid-State Circuits Conference*, pp. 415–418, 2011
2. D. Raiteri, F. Torricelli, A.H.M.V. Roermund, E. Cantatore, Design of a voltage-controlled oscillator based on organic TFTs. *International Conference on Organic Electronics*, 2012
3. G. Horowitz, W. Kalb, M. Mottaghi, P. Lang, Structure-performance relationship in penta-cene-based thin-film transistors. *IEEE International Symposium on Industrial Electronics*, pp. 1409–1411, 2004
4. A. Ulman, Formation and structure of self-assembled monolayers. Chem. Rev. **96**(4), 1533–1554 (1996)
5. F. Torricelli et al., Transport physics and device modeling of zinc oxide thin-film transistors Part I: Long-channel devices. IEEE Trans. Electron Devices **58**(8), 2610–2619 (2011)
6. F. Torricelli, Charge transport in organic and disordered semiconductor materials and devices, Ph.D. dissertation (2010)
7. M. Vissenberg, M. Matters, Theory of the field-effect mobility in amorphous organic transistors. Phys. Rev. B: Condens. Matter **57**(20), 964–967 (1998)
8. C. Tanase, E.J. Meijer, P.W.M. Blom, D.M.D. Leeuw, Unification of the hole transport in polymeric field–effect transistors and light-emitting diodes. Phys. Rev. Lett. **91**(21), 216601 (2003)
9. F. Torricelli, Z.M. Kovacs-Vajna, L. Colalongo, A charge-based OTFT model for circuit simulation. IEEE Trans. Electron Devices **56**(1), 20–30 (2009)
10. K. Myny et al., Unipolar organic transistor circuits made robust by dual-gate technology. IEEE J. Solid-State Circuits **46**(11), 1223–1230 (2011)
11. G.H. Gelinck, E. van Veenendaal, R. Coehoorn, Dual-gate organic thin-film transistors. Appl. Phys. Lett. **87**(7), 073508 (2005)
12. A. Cerdeira, M. Estrada, R. García, A. Ortiz-Conde, F.J.G. Sánchez, New procedure for the extraction of basic a-Si:H TFT model parameters in the linear and saturation regions. Solid-State Electron **45**, 1077–1080 (2001)
13. R.H. Crawford, *MOSFET in Circuit Design: Metal–Oxide-Semiconductor Field–Effect Transistors for Discrete and Integrated-Circuit Technology* (McGraw-Hill, New York, 1967)

Chapter 5
Sensor Frontend Architecture

In this chapter, the three main functional blocks in a smart sensor (analog signal conditioning, analog to digital conversion, and logic) are analyzed. It will be shown that most of the system, architectural, and circuital choices must be based on technology constraints and not on the target application, as it usually happens in standard technologies. First the analog signal conditioning chain is considered from a system point of view, choosing an architecture suitable for our dual-gate p-type only TFT technology. A deeper insight into analog interface limitations and solutions will be given in Chap. 6. Next, the reasons for choosing an integrating ADC are discussed. In Chap. 7, experimental data will confirm that this approach performs very well when compared to prior art solutions. Eventually, an extensive analysis on the state of the art in digital logic styles is carried on, and the main objectives for a new logic style which is robust to TFT variability are drawn. In Chap. 8, this new logic will be presented in detail and characterized with extensive statistical measurements.

5.1 Analog Signal Conditioning

In smart sensor applications, the analog signal coming from the sensor needs to be converted and sent to a base station where, taking advantage of higher computational power, data can be processed to extract meaningful information. Conversion to a digital format that enables robust radio transmission to the base station is essential to enable a practically usable system. Before doing so, the analog signal needs to be amplified and filtered to improve, or at least preserve, the signal to noise ratio (SNR) present at the sensor output, and to adjust the signal to the dynamic range (DR) of the analog-to-digital converter. This is called signal conditioning (Fig. 5.1).

Ideally, the transfer function of the whole conditioning chain is independent of the ordering chosen for the main functional blocks (i.e.: amplification, filtering, and sampling), but in practice a suitable order in the actual implementation must be chosen, taking into account the specifications and the limitations arising from the

© Springer International Publishing Switzerland 2015
D. Raiteri et al., *Circuit Design on Plastic Foils,* Analog Circuits and Signal Processing,
DOI 10.1007/978-3-319-11427-9_5

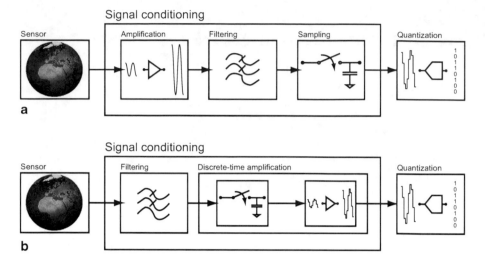

Fig. 5.1 Two possible signal conditioning chains

chosen technology. In sensor applications, typically the signal is first amplified and then filtered, sampled and quantized (Fig. 5.1a). In this way, the noise introduced by the electronics after the low-noise amplifier affects the SNR at the input of the signal conditioning chain only in a negligible way.

However, also other factors play a role in the quality of the analog signal processing and of the analog-to-digital conversion. In large-area electronics, for instance, it is very difficult to implement continuous-time amplifiers with sufficient gain and linearity: indeed linear resistors are typically not available, the TFTs have low intrinsic gain, and often only normally-on p-type transistors are available in a given technology.

In order to achieve enough gain to enable negative feedback circuits, the cascade of many amplifiers should be exploited. However, in unipolar technologies it is very cumbersome to match the DC output of one stage with the DC input of the following stage, resulting in a small gain improvement with increasing number of stages (e.g. from single to three stage amplifier respectively 12 dB, 20 dB, and 23 dB [1–3]). Each stage, however, introduces additional singularities in the transfer, with a detrimental effect on the stability of the circuit when a feedback is applied. For this reason, nested compensations would be mandatory causing a strong increase in the circuit complexity and a drop in circuit reliability. The DC level issue can be solved by exploiting complementary technologies. However, the large TFT threshold voltage often experienced in this kind of processes demands supply voltages as large as 50 V, while the large parasitics typical of these technologies limit the gain bandwidth product in amplifiers to a few tens of Hz [4].

In Chap. 6, is demonstrated a discrete-time amplifier which is based on a new device acting as parametric capacitor. This solution enables in our technology a gain of 20 dB, without detrimental effect on speed, and it is suitable for a larger input common mode range. This is possible because the amplification is achieved without charge transfer between different capacitors, but just by changing the value of the capacitor itself. As the gain of the parametric amplifier is only related to two capacitance values, the linearity is intrinsically guaranteed for any input voltage.

This interesting approach to improve the gain with negligible loss in speed and linearity is intrinsically discrete-time, and thus it can be implemented only after sampling, as will be discussed in more detail in Chap. 6. For this reason, a continuous time filtering is mandatory at the input of the conditioning chain, to avoid aliasing effects. Therefore, in this work the signal processing chain schematically shown in Fig. 5.1b is proposed.

In Chap. 6, for each function in Fig. 5.1b, a circuit design specifically adapted to our double-gate unipolar technology is proposed. On the one hand, our designs improve the robustness of the circuits against hard/soft faults and mismatch, favoring circuit simplicity and compactness. On the other hand, electrical tunability is proposed as a solution to counteract process parameter variations and aging.

The anti-alias filter is embodied by a $G_m C$ filter that exploits a tunable transconductor made of only five double-gate TFTs. The transconductance of the circuit is tunable over one order of magnitude providing resilience to process parameter variations and control on the cut-off frequency of the filter. Unfortunately, the solution proposed is open loop (due to the difficulty to create large gain and to use feedback stabilization in our technology). As the filter in our discrete-time implementation of the signal conditioning chain is preceding the signal amplification, linearity issues due to the amplitude of the signal should be negligible. It is true, however, that the DC level of the signals that can be handled by the proposed $G_m C$ filter is limited and signals having a DC component close to the ground or to the power supply level must be avoided. All details of the proposed $G_m C$ filter are discussed in Chap. 6.

The filtered signal is then sampled on the parametric capacitor in the first synchronous phase and amplified in the second phase, as will be discussed in more detail in Chap. 6. Then the discrete-time analog signal can be fed to the quantizer which converts the analog sample into a digital word.

5.2   Data Conversion

The design of data converters represents a delicate topic in any manufacturing technology due to their mixed-signal nature. Indeed this kind of circuits links different signal domains, converting a continuous-time and continuous-amplitude signal into a discrete-time quantized digital word (analog to digital converter—ADC), or vice versa (digital to analog converter—DAC).

In order to preserve the signal to noise and distortion ratio (SNDR) of the input signal, the converter should satisfy specific requirements in terms of resolution, linearity and speed. According to the application addressed and to the properties of the process in use, different architectural and circuit solutions can be chosen to favor one requirement or the other.

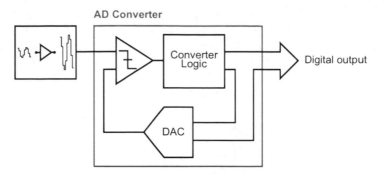

Fig. 5.2 Building block schematic of an ADC based on a DA in the feedback

In the field of flexible electronics, speed is not an essential issue since the quantities that must be sensed in these applications are most often quasi-static (temperature, chemical levels, large-area strains, etc). For this reason, the few examples of ADC and DAC in the state of the art focus on resolution and linearity.

A large category of ADCs are based on the comparison between the output of a DAC and the analog input, according to the general architecture shown in Fig. 5.2. For this class of ADCs, resolution and linearity are limited by the characteristics of the DAC used. In the DAC, a suitable reference is usually divided in smaller parts using matched unit elements (passives, like resistors and capacitors, but also actives, like TFTs). Due to the poor matching offered by the available passives and especially by the TFTs in large-area technologies manufactured at low temperature, ADCs built following this approach led, at the state of the art, to a maximum integral non-linearity (INL) of 2.6 LSB at 6 bit resolution level [5]. In particular, a 6 bit resolution DAC exploiting a C-2C approach [5] or a current-steering topology [6], have been demonstrated. In the case of the C-2C DAC, the converter was also used to realize a successive approximation register (SAR) ADC [5]. The non-linearity before calibration was larger than 3 LSB, and calibration was performed using external logic controlling an additional integrated 2 bit thermometric coded DAC. In this way, a maximum INL of 0.6 LSB @ 10 Hz and 1.5 LSB @ 100 Hz is achieved. More recently also an ADC based on an integrated, printed R-2R DAC has been demonstrated [7]. The DAC ensured good linearity (0.4 LSB linearity at 7 bit resolution level in the most recent reports [8], but it was limited to only 4 bit resolution due to the low level of circuit complexity that can be achieved with acceptable yield in the state-of-the-art complementary organic technology used for this design.

In the architectures that can be described according to Fig. 5.2, the ADC resolution is equal to the resolution of the DAC. A well-known method to increase the ADC resolution far beyond the resolution provided by the DAC is the use of oversampling and noise shaping, exploiting the feedback architecture schematically depicted in Fig. 5.3, which is called a $\Delta\Sigma$ modulator. This converter topology performs well when OTAs with large gain-bandwidth product and good DC gain (typically above 40 dB) are available, to create a loop filter with sufficient gain at the (over) sampling frequency.

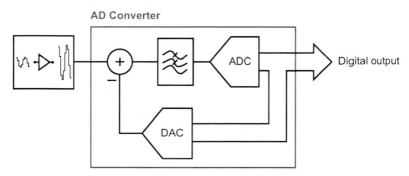

Fig. 5.3 Building block schematic of a first order $\Delta\Sigma$ modulator used as an AD converter

Unfortunately, this is difficult to achieve using large-area electronics on foil. As a consequence of the small GBW available from OTAs manufactured with OTFTs (even when using complex multi-stage amplifiers [1], the oversampling ratio is low, limiting the effectiveness of the $\Delta\Sigma$ approach. On top of this, the filter order is limited, because of the maximum circuit complexity that can be achieved with reasonable yield, limiting even more the advantage of using a $\Delta\Sigma$ structure. The equivalent number of bits obtained by the only example available in literature of a $\Delta\Sigma$ modulator made with OTFTs on foil was thus slightly higher than 4 bits [1].

A special case of DAC-based ADCs is represented by the dual-slope integrating ADC (Fig. 5.4). This topology is intrinsically robust to mismatch, since the conversion is provided as a ratio in the time domain, based on a stable reference in the time domain, the clock period, T_{clk}, rather than in the amplitude domain. During a first phase, the analog signal to be converted is integrated for a fixed number of clock periods. In the second phase, the integration of a reference analog signal is subtracted from the final value obtained in the first phase, and the number of clock cycles needed to zero the output of the integrator (N in Fig. 5.4) provides (together with the number of cycles in the first counting phase) the converted data. Following this approach, the matching problems in the integrator can be cancelled out and all the design effort can be focused on the linearity of the integration process.

A similar approach was followed in this work, further aiming to a reduction of the complexity of the converter and the area occupied. The solution proposed is a VCO-based ADC, where the analog signal is converted in a proportional frequency by a linear VCO. The integration of this output in a counter for given, fixed time, returns a quantized value proportional to the signal to be converted. This ADC requires only two building blocks: a VCO and a digital counter, moreover, the converter is not affected by TFT mismatch (see Chap. 7). However, it requires a stable time source, i.e. an integration time which is constant over the single measurement interval. This is the only constraints imposed by the system if the gain and offset correction approach suggested in Chap. 7 is followed. Under this approach, two integration times are required to generate the reference and one to convert the signal. If the stability of the time source is known to be sufficient over a long enough time, more consecutive signal conversions can be performed with only one reference generation.

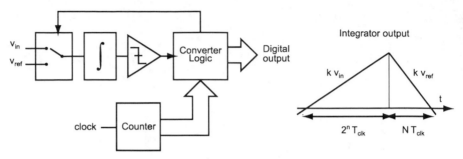

Fig. 5.4 Building block schematic of a dual-slope integrating ADC and time behavior of the integrator output

The proposed ADC stands out for its simplicity. Indeed, the reduced complexity allows a much more compact design compared to the state of the art ADCs, and can improve the linearity beyond state of the art performance even without calibration. A more detailed discussion on this data converter will be given in Chap. 7.

5.3 Robust Digital Blocks

Digital circuits are important in almost every integrated circuit, addressing the most diverse applications. In some cases, they implement the core functionality of the circuit (e.g. finite-state machines, CPUs, ALUs, code generators, shift registers, digital filters, and many more basic building blocks), while sometimes they assist analog circuits (as it is the case of the VCO-based ADC presented in Chap. 7) or they implement control and synchronization tasks.

Unfortunately, the same characteristics of large-area technologies that pose strong limitations on the performance of analog signal conditioning and data conversion functions, cause poor static performance in digital circuits, typically resulting in low gain and limited noise margin in the logic gates. For this reason, alternative logic styles must be developed to ensure better static performance and higher robustness by exploiting suitable circuit techniques.

The poor static performance of large-area digital electronics on foil eventually results in bad yield. Indeed, each logic gate in a digital circuit should ensure a sufficient noise margin, if the complete digital circuit must have a high probability to operate correctly. However, this may not be the case due to the combination of large TFT variability and small noise margin that is observed in unipolar logic gates for large-area low-temperature technologies.

Many different definitions of noise margin are available in literature [9] : for all of them however, both a large gain and a symmetric input–output characteristic are required to achieve high noise margin. In this book, is adopted the Maximum Equal Criterion (MEC) that defines the noise margin as the side of the maximum square that can be inscribed between the static input–output characteristic and its mirrored version.

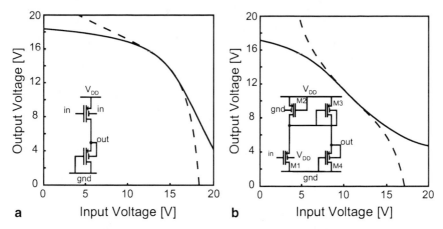

Fig. 5.5 Schematic and transfer characteristic of a diode loaded inverter **a** without and **b** with level shifter. *Dashed lines* show the mirrored characteristic. (© 2014 IEEE. Reprinted, with permission, from Raiteri et al. [13])

Table 5.1 Benchmark among state-of-the-art logic styles applied to similar technologies

	Diode load[a]		Zero-Vgs load[a]		Pseudo CMOS	Dual-gate enhanced		PLS[a]
	w/o LS[b]	w LS[b]	w/o LS[b]	w LS[b]	–	Diode	Zero-Vgs	–
Supply	20 V	20 V	20 V	20 V	20 V	20 V	20 V	20 V
NM	–	–	3 V	2.8 V	2.5 V	1.4 V	6 V	8.2 V
Gain	6.4 dB	1.6 dB	13 dB	11.6 dB	15.6 dB	6 dB	20 dB	76 dB
V_{trip}	15 V	10.5 V	13 V	6 V	7 V	10 V	10 V	10 V
V_{high}	18.6 V	17.5 V	19 V	17.7 V	19.9 V	16 V	18 V	19.5 V
V_{low}	4.2 V	4.7 V	0.1 V	0.4 V	0.1 V	0.1 V	0.1 V	0.1 V
L_{min}	5 µm	5 µm	5 µm	5 µm	5 µm	5 µm	5 µm	5 µm

[a] Fabricated in the same double-gate technology
[b] Level shifter

A simple way to design an inverter in a p-type-only technology is to use a common-source TFT for the pull-up and a TFT in diode configuration to implement the pull-down (Fig. 5.5a, inset). This inverter is rather fast, as both the pull-up and the pull-down actions are performed by strongly-on transistors. However, due to the small output resistance (in the order of 1/gm) the maximum gain of this inverter implemented in our technology is just about two. Due to the normally-on nature of our TFTs, the trip point is located around 3/4 V_{DD}. This fact, in combination with the small gain, results in a noise margin (NM) close to 0 V (Fig. 5.5a and Table 5.1). For this reason, a level shifter in front of the diode load inverter is often mandatory to make the transfer characteristic of the inverter more symmetrical and increase the noise margin [10].

The experimental transfer characteristics of two diode load inverters manufactured in our technology without and with level shifter are shown, together with their

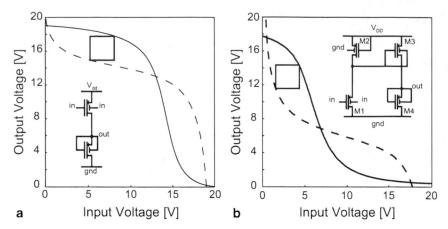

Fig. 5.6 Schematic and transfer characteristic of a Zero-Vgs loaded inverter **a** without and **b** with level shifter. *Dashed lines* show the mirrored characteristic. (© 2014 IEEE. Reprinted, with permission, from Raiteri et al. [13])

schematics, in Figs. 5.5a, b. Due to the gain lower than 6.4 dB, the measured noise margin is in both cases zero.

A first solution to improve the static noise margin of the previous inverter topologies is to replace the diode load with a Zero-Vgs connected TFT working in saturation (Fig. 5.6a). The increased output resistance and thus the larger gain improves the noise margin, but the time response is slower due to the weak pull-down action. For the Zero-Vgs load inverters with and without level shifter (Fig. 5.6a, b), the measured gain was respectively of 11.6 dB and 13 dB. The maximum noise margin decreased from 3 V to 2.8 V when using the level shifter, due to the excessive shift provided by this stage, which had in practice different TFT parameters from the ones used to simulate the design (Fig. 5.6 and Table 5.1).

Despite this problem, that was due to excessive variability, the use of a level shifter appears an effective solution to centre the inverter's characteristic within the input range. This is especially true if the shift amount can be controlled to cope with the actual TFT parameters.

Another interesting approach to design unipolar logic circuits exploits a two-stage inverter (Fig. 5.7a) [11], also called a pseudo-CMOS inverter. The output transistor M4 of the second stage is driven by a diode load inverter (M1–M2) supplied by V_{DD} and V_{SS} (which is below the ground voltage, gnd). In this way, when the input is low and M3 pulls the output up to the high logic state, the internal voltage V_i disables the pull-down action of M4, pulling its gate up. On the other hand, when the input is high, V_i goes low and M1 brings the gate of M4 below gnd, pulling effectively the output node to ground.

Since both pull-up and pull-down actions are performed in this inverter by a strongly-on TFT, both good dynamic characteristics and a reasonable gain (15.6 dB [11]) can be achieved. The main disadvantage is that to achieve a good noise margin (trip point at 7 V and NM = 2.5 V [11] this configuration needs a large negative

Fig. 5.7 **a** Schematic of a Pseudo-CMOS inverter. **b** Schematic of a dual-gate enhanced inverter with diode (*left*) and Zero-Vgs (*right*) load. (© 2014 IEEE. Reprinted, with permission, from Raiteri et al. [13])

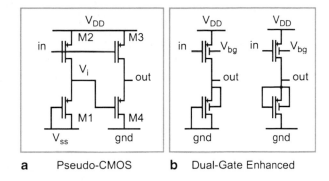

supply ($V_{ss} = -24$ V), hence the design of a circuit able to automatically control the value of V_{ss} to adjust the trip point of the inverter is not trivial and must include switching power-management circuits.

The last noteworthy logic style used for TFTs on foil is the dual-gate enhanced inverter [12] that exploits the double-gate technology presented in Chap. 4 to shift the transfer characteristics on the input–output plane. The load TFT can be either diode or Zero-Vgs connected (Fig. 5.7b). The simple idea behind the dual-gate enhanced logic is to use the top-gate bias V_{bg} to vary the threshold voltage of the pull-up TFT, making it a normally-off transistor, and thus shifting the inverter transfer characteristic to the left, towards the center of the input range. This approach provides outstanding results in terms of symmetry of the transfer characteristics: choosing V_{bg} suitably, it is indeed possible to shift the trip point exactly to the center of the input range, improving dramatically the noise margin (NM~1.4 V for diode load and NM~6 V for Zero-Vgs load [12], see Table 5.1). Unfortunately, due to the weak effect of the top-gate voltage on the TFT threshold, large values for V_{bg}, typically above V_{DD}, are needed. Moreover the top-gate bias reduces the gate overdrive of the pull-up TFT and hence the speed of the pull-up action.

Reviewing the most popular state-of-the-art logic styles used with TFT processed at low temperature, it was shown that high gain and a symmetrical transfer characteristic are required to achieve large noise margin in an inverter. Moreover, in a large digital circuit, each logic gate should have sufficient noise margin to guarantee robust functionality and enable good yield in spite of the variability of TFT parameters. For this reason, a circuit technique allowing post-fabrication tuning of the trip point to maximize the noise margin would be very beneficial to manufacture on foil digital circuits of some complexity and still acceptable yield. The solutions known from prior art and discussed in this section require tuning voltages outside the supply range, which are cumbersome to generate in typical large-area applications, as they require switching circuits and large, high-quality passives. In fact, the pseudo-CMOS inverter requires a negative supply as large as $V_{ss} = -24$ V (with $V_{DD} = 20$ V), while, according to Eq. (4.10), to shift the dual-gate enhanced inverter characteristic of $\Delta V_{trip} = 8$ V, and center the input–output characteristic (Fig. 5.7 in [12] in the supply range, the top-gate of the driver should be biased at $V_{bg} = V_{DD} + \Delta V_{trip}/k \sim 52$ V.

To avoid the need for additional complexity and allow self-correction on chip, a new logic style is proposed that exploits positive feedback to enable tuning voltages within the supply rails while ensuring higher gain and larger tunability of the input–output characteristic than state of the art solutions. The details of this new logic style will be described in Chap. 8.

References

1. H. Marien, M.S.J. Steyaert, E. van Veenendaal, P. Heremans, A fully integrated ΔΣ ADC in organic thin-film transistor technology on flexible plastic foil. IEEE J. Solid-State Circuits. 46(1), 276–284 (2011)
2. H. Marien, M. Steyaert, N.V. Aerle, P. Heremans, A mixed-signal organic 1 kHz comparator with low VT sensitivity on flexible plastic substrate. IEEE European Solid-State Circuit Conference (2009), pp. 120–123
3. H. Marien, M.S.J. Steyaert, E.v. Veenendaal, P. Heremans, Analog building blocks for organic smart sensor systems in organic thin-film transistor technology on flexible plastic foil. IEEE J. Solid-State Circuits. 47(7), 1712–1720 (2012)
4. G. Maiellaro et al., High-gain operational transconductance amplifiers in a printed complementary organic TFT technology on flexible foil. IEEE Trans. Circuits Syst. I Reg. Pap. 60(12), 3117–3125 (2013)
5. W. Xiong, U. Zschieschang, H. Klauk, B. Murmann, A 3 V 6b successive-approximation ADC using complementary organic thin-film transistors on glass. IEEE International Solid-State Circuits Conference (2010), pp. 47–49
6. T. Zaki et al., A 3.3 V 6-bit 100 kS/s current-steering digital-to-analog converter using organic p-type thin-film transistors on glass. IEEE J. Solid-State Circuits. 47(1), 292–300 (2012)
7. S. Abdinia et al., A 4b ADC manufactured in a fully-printed organic complementary technology including resistors. IEEE International Solid-State Circuits Conference (2013) pp. 106–107
8. S. Abdinia, S. Jacob, R. Coppard, A. H. M. v. Roermund, E. Cantatore, A printed DAC achieving 0.4 LSB maximum INL at 7b resolution level, in ICT.OPEN2013 (2013)
9. J.R. Hauser, Noise margin criteria for digital logic circuits. IEEE Trans. Educ. 36(4), 363–368 (1993)
10. E. Cantatore et al., A 13.56 MHz RFID system based on organic transponders. IEEE J. Solid-State Circuits. 42(1), 82–92 (2007)
11. T. C. Huang, K. T. Cheng, Design for low power and reliable flexible electronics: self-tunable cell-library design. J. Display Technol. 5(6), 206–215 (2009)
12. K. Myny et al., Unipolar organic transistor circuits made robust by dual-gate technology. IEEE J. Solid-State Circuits. 46(11), 1223–1230 (2011)
13. D. Raiteri, P. v. Lieshout, A. v. Roermund, E. Cantatore, Positive-feedback level shifter logic for large-area electronics. IEEE J. Solid-State Circuits. 49(2), 524–535 (2014)

Chapter 6
Circuit Design for Analog Signal Conditioning

In Chap. 5, have been explained the architectural choices for the design of the signal conditioning chain for smart sensors manufactured with TFTs on foil. In this chapter, each of the blocks in the signal conditioning chain is discussed separately, providing a deeper insight in the technology-aware circuit design techniques that have been developed to improve robustness and performance. Following the signal through the front-end path, the low-pass filter is introduced first, and then the continuous-time and the discrete-time amplifiers. Parts of this chapter have been published in [1–4].

6.1 A Filter Based on a Tunable Transconductor

Emerging technologies on foils still have huge room for improvement in terms of transistor performance, process uniformity, matching and stability. Indeed, the reliability of the process and the performance of TFTs are much inferior to what analog designers are used to in standard silicon technologies. In light of this consideration, the continuous-time low-pass filter, which is needed at the input of the signal conditioning chain for the reasons explained in Chap. 5, was designed targeting low circuit complexity, and enabling electric tuning to provide resilience to the large process variation typical of large-area electronics.

6.1.1 The Proposed G_mC Filter

A G_mC filter is a continuous-time low-pass filter that, in its simplest implementation, employs only a transconductor and a capacitive load (Fig. 6.1a).

The small-signal output current i_{out} is proportional to the small-signal input voltage through the transconductance G_m of the transconductor (not to be confused with the TFT transconductance g_m). The output current is afterwards integrated by the loading capacitor and thus converted in the output voltage v_{out} (signals will be

© Springer International Publishing Switzerland 2015   53
D. Raiteri et al., *Circuit Design on Plastic Foils,* Analog Circuits and Signal Processing,
DOI 10.1007/978-3-319-11427-9_6

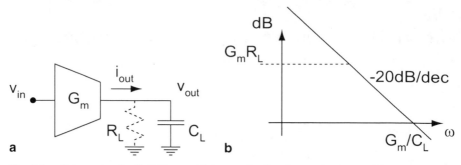

Fig. 6.1 **a** Schematic of a G_mC filter and **b** its transfer function v_{out}/v_{in}. In a real implementation, the output resistance limits the DC gain of the circuit

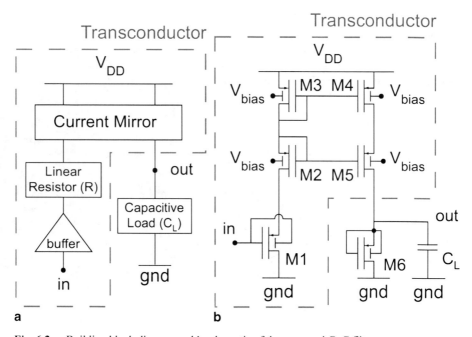

Fig. 6.2 **a** Building block diagram and **b** schematic of the proposed G_mC filter

expressed as $v_X = V_X + v_x$, that is the sum of a DC component V_X and a small signal v_x). Due to the integration operation, the amplitude Bode plot of the voltage transfer v_{out}/v_{in} (Fig. 6.1b) has a -20 dB/dec slope and crosses the 0 dB amplification line at a frequency proportional to the ratio between transconductance G_m and the loading capacitance C_L. Moreover, at the output node there is always a resistive load R_L (typically due to the channel length modulation of the transconductor output devices) which limits the DC gain at low frequencies.

The block diagram of the G_mC filter proposed in this book is shown in Fig. 6.2a. A voltage buffer is used to apply the input voltage to a linear resistor R. In the case of an ideal buffer and current mirror (which have zero output and input impedance, respectively), the whole input voltage drops over the linear resistor and the current generated is mirrored to the output port using the output branch of the mirror, which is loaded with the capacitor. Therefore, the ideal transconductance of this circuit is:

$$G_{m,ideal} = \frac{1}{R}.$$ (6.1)

In the transistor level implementation of the circuit (Fig. 6.2b), transistor M1 implements the buffer, the current mirror is made of M3, M4 and M5, and the linear resistor is embodied by M2. Transistor M6, connected in parallel to the load capacitor C_L, is required to bias correctly the filter and, in first approximation, its large output resistance does not affect the small-signal behavior of the circuit.

6.1.2   Transconductor Analysis

The core of the transconductor is the linear resistor R, which converts the voltage applied by the buffer in a proportional current (Fig. 6.2a). In our technology, like in almost all large-area electronics, passive linear resistors are not available, hence a transistor has to be used for this purpose. Two possible choices are available: a TFT exploited in its linear region or in saturation. The G_mC filter proposed here targets low-frequency applications, hence a low transconductance and a low cut-off frequency are pursued by selecting the second option, i.e. M2 in saturation. Indeed, since the drain-source resistance of the TFT determines the actual transconductance G_m of the transconductor, the larger drain-source resistance obtained when M2 is used in saturation causes a smaller transconductance and allows, for a given cut-off frequency, a smaller load capacitance. Also, from the linearity point of view, the current with respect to the (drain-source) voltage is more linear in the saturation region than in the triode one.

In order to achieve saturation already for a small drain-source voltage, source and gate of transistor M2 are connected together (as before, this configuration will be referred to with Zero-Vgs connection, i.e. $v_{GS} = 0$ V). This is a viable solution in organic unipolar technologies, since OTFTs are typically normally-on devices, as underlined in Chap. 4. If Zero-Vgs connected, the TFT enters the saturation region when the applied drain-source voltage is larger than the threshold voltage of the transistor. In our p-type double-gate technology, the threshold voltage of the device can be controlled by means of the voltage applied to the top-gate. In fact, making the voltage applied to the top-gate more and more positive shifts the threshold to more negative values, hence it reduces the channel current, increases the output resistance of the OTFT, and extends the width of the saturation region (Fig. 6.3). Due to the actual implementation of the linear resistor, its resistance R in Eq. (6.1) can be replaced with the small-signal output resistance of M2 in saturation, i.e. $r_{0,2}$. Moreover,

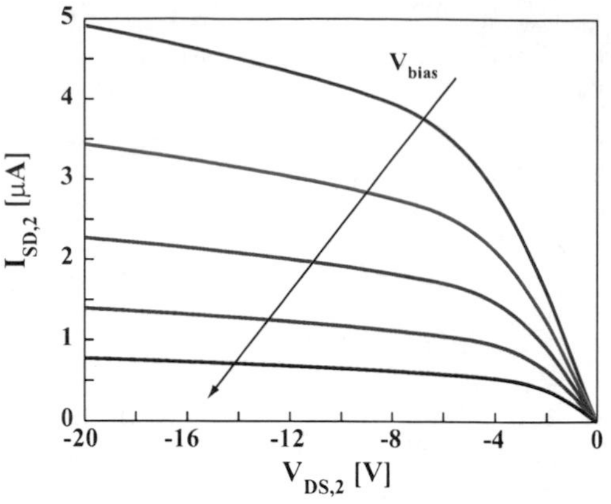

Fig. 6.3 Simulated output
characteristic of a Zero-Vgs
connected p-type OTFT
for different top-gate bias
$V_{bias} = -20, -15, -10, -5,$
0 V

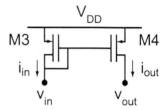

Fig. 6.4 Basic p-type-only
current mirror schematic

thanks to the influence of the control voltage V_{bias} on the output resistance the
$G_{m, ideal}$ of the transconductor can be effectively tuned.

In a real implementation of the circuit, however, the actual transconductance
G_m is smaller than $G_{m, ideal}$. The main reason for this reduction is the actual small-
signal mirroring factor which is smaller than 1 even if M3 and M4 have the same
W/L ratio, the same voltage bias, and the same current bias. This peculiar behavior
can be explained considering at first the simplest current mirror topology shown
in Fig. 6.4. Since transistor M3 is normally-on, connecting together drain and gate
always forces a linear regime (and not saturation). Therefore a small-signal applied
to the terminal v_{in} causes a current i_{in} made of two contributions: the first one due to
the usual gate transconductance, the second due to the drain conductance. In standard technologies the drain conductance depends on the channel length modulation
and it is typically negligible, while in our case it depends on the (much larger) triode
conductance. This contribution however is not mirrored to the output branch, since
the drain of the output transistor is biased at a constant voltage V_{OUT} (i.e. $v_{out} = 0$ V).
Under this assumption, the small-signal output current is always smaller than the
input one.

This circuit can be analyzed using the TFT equations discussed in Chap. 4. For
the sake of simplicity, the analysis is made for an n-type current mirror, since the
equations derived in Chap. 4 are also written for n-type transistors. Applying a small
input signal v_{in} and considering our current model (Eq. (4.4)), the small-signal input

current i_{in} flowing through the sink transistor M3 can be expressed as a function of the gate transconductance ($\delta I_{DS}/\delta V_{GS}$) and the drain conductance ($\delta I_{DS}/\delta V_{DS}$):

$$i_{in} = \frac{\delta I_{DS}}{\delta V_{GS}} v_{gs} + \frac{\delta I_{DS}}{\delta V_{DS}} v_{ds} = \frac{W_3}{L_3} \beta \gamma \left[\left(V_{G,3} - V_{S,3} - V_{FB} \right)^{\gamma-1} - \left(V_{G,3} - V_{D,3} - V_{FB} \right)^{\gamma-1} \right] v_{in}$$

$$+ \frac{W_3}{L_3} \beta \gamma \left(V_{G,3} - V_{D,3} - V_{FB} \right)^{\gamma-1} v_{in} \qquad (6.2)$$

where $V_{G,X}$, $V_{S,X}$ and $V_{D,X}$ are the DC gate, source and drain voltage of the transistor MX, and V_{FB} is the flat-band voltage.

On the other hand, the small-signal output current of the mirror only depends on the voltage variation at the gate of M4, since its drain voltage is assumed to be constant (i.e. $v_{out} = 0$ V), hence it reads:

$$i_{out} = \frac{W_4}{L_4} \beta \gamma \left[\left(V_{G,4} - V_{S,4} - V_{FB} \right)^{\gamma-1} - \left(V_{G,4} - V_{D,4} - V_{FB} \right)^{\gamma-1} \right] v_{in}. \qquad (6.3)$$

From the ratio between these two equations, the mirroring factor T can be expressed as:

$$T = \frac{i_{out}}{i_{in}} = \frac{\frac{W_4}{L_4} \beta \gamma \left[\left(V_{G,4} - V_{S,4} - V_{FB} \right)^{\gamma-1} - \left(V_{G,4} - V_{D,4} - V_{FB} \right)^{\gamma-1} \right]}{\frac{W_3}{L_3} \beta \gamma \left(V_{G,3} - V_{S,3} - V_{FB} \right)^{\gamma-1}}. \qquad (6.4)$$

Considering $V_S = V_{S,3} = V_{S,4}$, $V_G = V_{G,3} = V_{G,4}$ and $V_D = V_{D,4}$, and assuming identical W and L for M3 and M4, Eq. (6.4) can be then simplified as:

$$T = 1 - \left(\frac{V_G - V_D - V_{FB}}{V_G - V_S - V_{FB}} \right)^{\gamma-1}. \qquad (6.5)$$

Equation (6.5) clearly shows that T is always smaller than 1 even when the voltage bias is the same for the input and the output nodes, i.e. $V_D = V_G$, and the bias currents are identical. Moreover, Eq. (6.5) holds for both saturation and linear regimes; in fact, in order to derive it no assumption was made on the working region of the two devices.

6.1.3   Transconductor Design

From the simplified expression of T, it can also be inferred that a mirroring factor close to unity is obtained when the drain voltage of the output transistor is biased around V_G. For this reason, in the final implementation of the current mirror, the additional transistor M5 (Fig. 6.2b), with the same dimensions of M2, was included

Fig. 6.5 Unloaded transconductor for intrinsic voltage gain evaluation

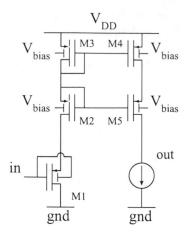

to keep the output TFT M4 in the linear region like M3. The transistor M5, however, acts as source follower with respect to the small-signal variation v_{in}. Therefore, the drain of M4 is not constant anymore and its drain conductance also contributes to the mirrored current. It is easy to derive that the mirroring factor T_{M5} in presence of M5 becomes:

$$T_{M5} = 1 - \frac{1}{1 + g_{m,5}/g_{m,4}} \left(\frac{V_G - V_D - V_{FB}}{V_G - V_S - V_{FB}} \right)^{\gamma-1}.$$

(6.6)

Indeed, the small-signal variation of the drain of M4 is given by the partition of the input voltage ($v_{S,2}$ in Fig. 6.5) between the channel resistance of M5 ($1/g_{m,5}$) and the channel resistance of M4 ($1/g_{m,4}$) which acts as a resistive source degeneration for M5.

In order to derive Eq. (6.6) the voltage drop on this resistor can be expressed as:

$$v_d = v_{in} \frac{\frac{1}{g_{m,4}}}{\frac{1}{g_{m,4}} + \frac{1}{g_{m,5}}} = v_{in} \frac{g_{m,5}}{g_{m,4} + g_{m,5}} = S v_{in}.$$

If we define

$$A = \frac{W_3}{L_3} \beta \gamma \left(V_G - V_S - V_{FB} \right)^{\gamma-1}$$

and

$$B = \frac{W_3}{L_3} \beta \gamma \left(V_G - V_D - V_{FB} \right)^{\gamma-1},$$

still considering that $V_S = V_{S,3} = V_{S,4}$, $V_G = V_{G,3} = V_{G,4}$ and $V_D = V_{D,4}$, Eqs. (6.2) and (6.3) can be expressed as:

$$i_{in} = \frac{\delta I_{DS}}{\delta V_{GS}} v_{gs} + \frac{\delta I_{DS}}{\delta V_{DS}} v_{ds} = A v_{in}$$

and

$$i_{out} = \frac{\delta I_{DS}}{\delta V_{GS}} v_{gs} + \frac{\delta I_{DS}}{\delta V_{DS}} v_{ds} = A v_{in} - B v_{in} + B S v_{in}.$$

From the ratio between these two currents, T_{M5} can be obtained as:

$$T_{M5} = \frac{A v_{in} - B v_{in} + B S v_{in}}{A v_{in}} = 1 - \frac{B}{A}(1 - S) = 1 - \frac{B}{A}\left(1 - \frac{g_{m,5}}{g_{m,4} + g_{m,5}}\right)$$

$$= 1 - \frac{B}{A}\frac{g_{m,4}}{g_{m,4} + g_{m,5}} = 1 - \frac{1}{1 + \frac{g_{m,5}}{g_{m,4}}}\left(\frac{V_G - V_D - V_{FB}}{V_G - V_S - V_{FB}}\right)^{\gamma - 1}$$

which is Eq. (6.6).

Besides imposing the right bias to the output device of the mirror and buffering a portion of the input voltage to the output, the transistor M5 also determines the output resistance of the transconductor. It is worth noting however that the presence of M5 does not increase the output resistance as the cascode configuration would suggest. Indeed, since the source degeneration of M5 is weak and the resulting gain of the local negative feedback is low, the output resistance of the transconductor is, at first order approximation, equal to the output resistance of M5.

Aware of the parameters that affect the transconductance and the output resistance of the circuit, interesting considerations can be drawn about the small-signal voltage gain of the transconductor. Indeed, when the output is loaded with an ideal current source (a condition that will be referred to as 'unloaded'), the transconductance and the output resistance are determined by the output resistance r_0 of the two OTFTs M2 and M5 respectively. Therefore, with reference to the voltages v_{in} and v_{out} defined in Fig. 6.5, the unloaded voltage gain reads:

$$G = \frac{v_{out}}{v_{in}} = T_{M5}\frac{r_{05}}{r_{02}}. \tag{6.7}$$

This equation shows that the gain mainly depends on the channel length modulation affecting the transistors M2 and M5. For this reason, in order to increase the voltage gain, it is possible to change the dimensions of M5 to decrease its channel length modulation with respect to the one of M2. Table 6.1 summarizes the results of different simulations where the channel width W and the channel length L of M5 have been scaled up by the same factor S.

The values of $R_{out} = v_{out}/i_{out}$, extrapolated from simulation data obtained for different channel lengths, show that increasing the channel length, the output resistance R_{out} raises and so does the gain G. This scaling however does not produce a proportional increase in the gain by a factor S, as it would be expected from

Table 6.1 Simulated gain of the circuit in Fig. 6.5 for different channel lengths (and constant W/L) of M5

S	W [mm]	L [μm]	G_m [nA/V]	R_{out} [MΩ]	Gain
1	1	5	4.55	228	1.03
2	2	10	4.51	491	2.21
4	4	20	3.8	927	3.52
8	8	40	2.9	1800	5.22

Eq. (6.7); in fact, increasing the channel length of M5 varies its saturation behavior with respect to M2, and larger gate-source and drain-source voltages are needed to compensate for the smaller channel length modulation. For this reason the drain voltage of M4 gets closer to V_{DD}, causing a drop of the mirroring factor T_{M5} and consequently of the overall transconductance G_m. This drop can also be seen in Eq. (6.6). Indeed, the drain voltage of the n-type equivalent of M4 will decrease for longer channel lengths of the output device M5, and the numerator of the fraction in Eq. (6.6) will increase. In the circuit implementation of the transconductor that has been fabricated, S was chosen equal to 1.

So far only the dimensioning of the symmetric devices in the input and output branch was considered, i.e. M2 compared to M5 and M3 with respect to M4. However, the dimensions of M2 also need to be carefully optimized accordingly to the dimensions of M3 and M1, in order to avoid excessive voltage drops over these devices. In this case, indeed, the small signal gain and the transconductance remain unchanged, but the DC input range for which the transfer characteristic can be considered linear is reduced.

Figure 6.6 depicts indeed three possible scenarios for different relative dimensions of the input branch devices. If M2 is much wider than M1 and M3, the voltage drops V_{GS1} and V_{GS3} will be large, decreasing the linearity for high input voltages and drastically reducing the linear input range (Fig. 6.6, blue line). On the other hand, if M3 is wider than M2 this would result in a waste of area, while a wide M1 would cause non-linearities for low input voltages (Fig. 6.6, green line). Indeed, for low inputs, the source of M1 would saturate to ground due to the positive threshold voltage. Hence the linear part of the characteristic would not start for $v_{IN} = 0$ V, but for $v_{IN} > V_{GS1}(I_{MAX})$. According to these considerations the final design adopts the same dimensions for all the devices of the input branch (Fig. 6.6, red line).

In line with these considerations, all the devices in the transconductor are dimensioned with equal channel widths and channel lengths. However, also the transistor threshold plays a role in the performance of the circuit. Hence, in order to keep the electrical matching among the different devices, the bias voltage V_{bias} is applied to all transistors M2, M3, M4 and M5.

6.1.4   Transconductor Realization

The proposed transconductor was realized and measured. The layout of the circuit is shown in Fig. 6.7.

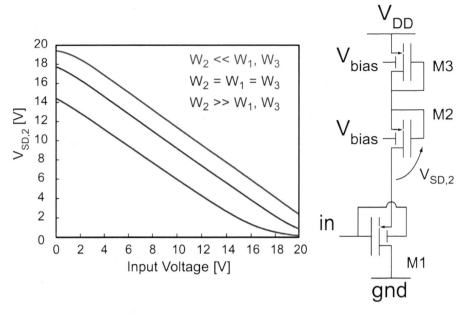

Fig. 6.6 Voltage drop on the transconductor device as a function of the input voltage for different choices of the transistor dimensions

Fig. 6.7 Layout of a tunable transconductor

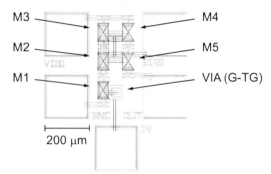

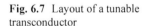

According to the discussion in Sect. 6.1.3, the TFTs in the circuit all have the same dimensions. In order to get a feeling of the real circuit dimensions, the width of the test pads is given. The transconductor occupies an area of about 200 μm × 400 μm.

6.1.5   *Transconductor Measurements and Simulations*

The circuit was designed to operate at 20 V supply, which is a relatively large value compared to the standard silicon technologies due to the thick dielectric layer. In order to get an insight in the influence of the control voltage, many different

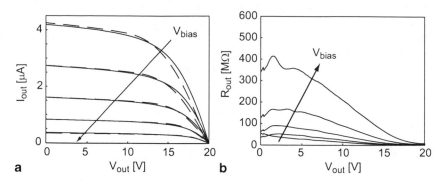

Fig. 6.8 **a** Measured (*continuous lines*) and simulated (*dashed lines*) output current as a function of the output voltage V_{out} for different values of $V_{bias}=0, 5, 10, 15, 20$ V ($V_{in}=5$ V) and **b** output resistance estimated from the measured data

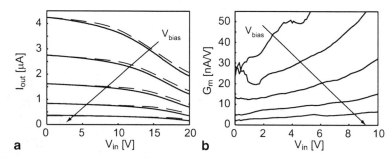

Fig. 6.9 **a** Measured (*continuous line*) and simulated (*dashed line*) output current as a function of the input voltage V_{in} for different values of $V_{bias}=0, 5, 10, 15, 20$ V ($V_{out}=5$ V) and **b** transconductance estimated from the measured data

measurements have been taken for $V_{bias}=0, 5, 10, 15, 20$ V. In Fig. 6.8 and 6.9, the step used for the independent variable was 100 mV for both the measurements and the simulations.

From the measurements of the output current obtained applying a constant voltage $V_{IN}=5$ V and sweeping v_{OUT} from ground to V_{DD} (Fig. 6.8a), the output resistance of the transconductor can be evaluated (Fig. 6.8b). As it has been shown already in Fig. 6.3, also in this case the control voltage V_{bias} can be used to modify the current and the output resistance of the transconductor. When this bias gets closer to ground, the output current raises and hence the output resistance drops. The maximum output current that was measured, exploiting a control bias between the rails, ranges from 4.098 µA for $V_{bias}=0$ V to 337 nA for $V_{bias}=20$ V. The relative variation is about one order of magnitude and the same variation can be seen also in the output resistance.

The transfer characteristic of the transconductor was also measured for the same values of V_{bias} (0, 5, 10, 15, 20 V), see Fig. 6.9a. The input node was swept from ground to V_{DD} while the output node was biased by an ideal voltage source at $V_{OUT}=5$ V. The resulting transconductance as a function of the input voltage v_{IN} is

Table 6.2 Measured transconductor parameters for different bias voltages

V_{bias} [V]	R_{out} [MΩ]	G_m [nA/V]	Gain
0	27	51	1.37
5	44	32	1.4
10	76	18	1.36
15	153	9.5	1.45
20	342	4.7	1.6

shown in Fig. 6.9b. Also in this case the output current increases when the control voltage approaches ground and so does the transconductance of the circuit: for instance, varying the control voltage from ground to V_{DD}, the minimum transconductance $G_{m,min}$ goes from 19 to 2 nA/V.

From Fig. 6.9 the influence of V_{bias} on the linearity of the circuit can also be evaluated. The higher the control voltage (i.e. lower current), the larger is the linear input range or, with the same input range, a higher linearity is achieved. Although in Fig. 6.9b the plots are less clean due to the low current and to the derivative operation required to extrapolate the transconductance, it is still possible to evaluate qualitatively the linearity of the transconductor for varying input bias by inspecting the shape of the curve. Indeed for larger V_{bias} (i.e. lower current), the small-signal transconductance G_m increases linearly with the bias point V_{IN} which results in a dominant second-order non-linearity in the $v_{in}\text{-}i_{out}$ characteristic. For lower V_{bias} (i.e. higher current), the transconductance G_m increases more than linearly raising the input bias V_{IN}, thus an higher V_{bias} allows more linear $v_{IN}\text{-}i_{OUT}$ characteristic.

The set of data in Figs. 6.8 and 6.9 (summarized in Table 6.2 for $V_{IN} = 5$ V and $V_{OUT} = 5$ V) also confirms that the unloaded gain of the circuit is almost independent on the bias voltage (and hence on V_T), since it depends on the ratio between the output resistances of the devices M2 and M5. The two devices have here the same W/L ratio and the same channel length hence the voltage gain is about one. The actual gain value is slightly higher than 1 because the output resistance of the circuit is not only due to the output resistance of M5, but also to the output resistance of M4. This contribution could be taken into account in Eq. (6.7) replacing $r_{0,5}$ with the complete output resistance $g_{m,5} r_{0,5} r_{0,4} + r_{0,5} + r_{0,4}$. Decreasing the output current (i.e. larger V_{bias}), the output resistance of M4 increases constantly, and this effect more than compensates the reduction in transconductance G_m due to the actual transfer factor T_{M5} and to the source follower M1.

Connecting together input and output nodes, a tunable resistor connected to V_{DD} is obtained. The measured current of this configuration is shown in Fig. 6.10 for different values of the control voltage V_{bias}.

6.1.6 Tunable Filter

After analyzing in depth the static performance of the transconductor, the behavior of this circuit can be shown when employed in a $G_m C$ filter, according to the schematic of Fig. 6.11.

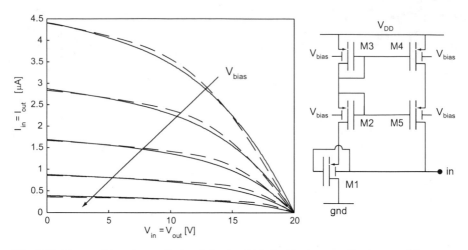

Fig. 6.10 Measured (*continuous line*) and simulated (*dashed line*) current flowing out of the transconductor with V_{in} connected to V_{out}, as a function of the input voltage V_{in} for different values of $V_{bias} = 0, 5, 10, 15, 20$ V

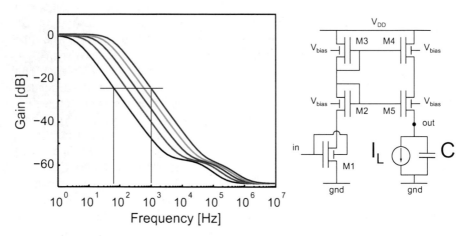

Fig. 6.11 Bode magnitude plot of the tunable transconductor. Varying the control voltage V_{bias} within the supply range, the first pole can be shifted over one decade of frequencies

The main feature of the proposed transconductor is the possibility of varying the transconductance of the circuit without affecting its unloaded voltage gain. For this reason, a filter can be implemented which has a bandwidth that is tunable over a decade of frequencies and, accordingly, the frequency response is shifted rigidly (Fig. 6.11).

The unity DC gain is retained, because the current source employed in the simulations of Fig. 6.11 has an infinite output resistance. If the load is implemented with a Zero-Vgs connected TFT (M6 in Fig. 6.12) to embody the current source, due to its finite output resistance, a reduction of the gain is observed (Fig. 6.12), while the gain-bandwidth product is preserved.

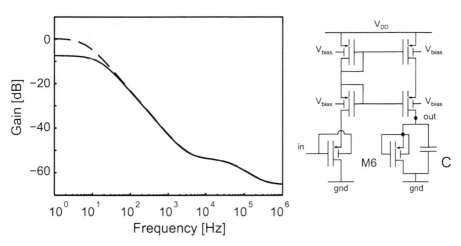

Fig. 6.12 Bode magnitude plot of the circuit with Zero-Vgs load (*continuous line*) and transfer function using an ideal current source instead of transistor M6 (*dashed line*)

6.2   Amplifiers

The most frequent building block in analog and mixed-signal design is probably the differential voltage amplifier, often called Operational Amplifier (OpAmp) since it can be used to implement the four arithmetical operations (addition, subtraction, multiplication and division) and many more functions. Indeed, since it provides negative-feedback systems with the needed high gain, this is a key element in integrators, differentiators, peak detectors, comparators, zero-crossing detectors, switched-capacitor circuits, sample and holds, and more.

6.2.1   A Continuous-time Amplifier

In large-area electronics, due to the many limitations posed by the technology, the design of high gain OpAmps is very cumbersome, and involves large circuit complexity that can be very detrimental for reliability.

On the one hand, due to the low intrinsic gain of TFTs, the use of multiple stages becomes mandatory to achieve high gain. On the other hand, the availability of only p-type TFTs causes two main issues: first each stage has a p-type input pair, thus suitable level shifters should be employed to match the input and the output voltage ranges, or AC coupling must be used; second, no self biasing load can be exploited, i.e. n-type current mirrors are not available. Moreover, telescopic solutions do not help, since cascoding can be applied only to the input pair.

The supply budget dedicated to each TFT in a stack is limited, and also with a supply voltage of about 20 V it is difficult to keep all transistors in saturation (remember that normally-on TFTs require much larger drain-source voltage to enter saturation than normally-off TFTs, typically larger than 6–8 V in our technology,

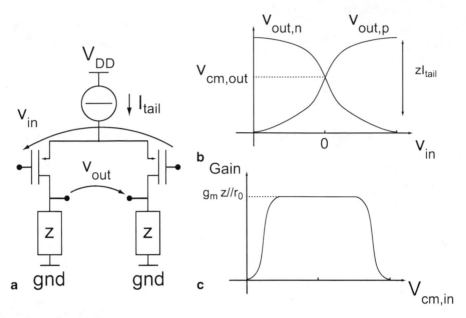

Fig. 6.13 **a** Schematic of a single-stage fully-differential amplifier, here **b** the gain between differential input and differential output also depends on **c** the common mode of the differential input

see Fig. 6.3) and linear resistors are not available. For this reason, even when only three TFTs are stacked between gnd and V_{DD}, like in fully-differential amplifiers exploiting Zero-Vgs active loads (Fig. 6.14b), the differential gain is very sensitive to input common mode and TFT biasing. Thus, compared to standard silicon technologies, the differential gain can decrease drastically with bias variations due to mismatch and/or common-mode shifts.

The Proposed Amplifier

Given these considerations, a simple continuous-time amplifier was chosen in this work. This structure can only achieve low gain, but ensures relatively fast response and high robustness. The circuit exploits a tail current source, a p-type input pair and active loads (Fig. 6.13a).

The gain of the amplifier is the ratio between differential output v_{out} (which is given by the difference between the positive and the negative outputs, $v_{out, p}$ and $v_{out, n}$ in Fig. 6.13b) and the differential input v_{in}. Other requirements, like the input and the output common-modes, need to be taken into account. Indeed this circuit works properly if the tail, the input and the load devices all work in saturation; however, the gain can drastically decrease if the input common mode $V_{CM, IN}$ approaches V_{DD} or gnd (Fig. 6.13c).

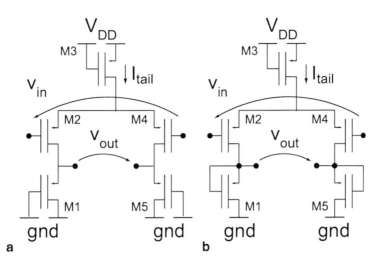

Fig. 6.14 Schematics of two fully-differential amplifiers exploiting **a** diode or **b** Zero-Vgs active load. Reprinted from D. Raiteri et al. [4], © 2014, with permission from Elsevier

Amplifier Analysis

The many limitations imposed by the technology restrict the design choices for a voltage amplifier. In our case, since the technology is p-only, the input pair can only be p-type (M2 and M4 in Fig. 6.14).

The tail current source should also be a p-type TFT and work in saturation, but unfortunately it cannot be the output transistor of a current mirror (as explained in Sect. 6.1.2). However, it can be easily embodied by a p-type TFT used in a Zero-Vgs configuration (M3 in Fig. 6.14) to provide high output resistance and thus a relatively supply-independent bias current I_{tail}.

Neither passive loads nor self biasing solutions are possible, due to the lack of resistors and complementary devices. Therefore the output loads can only be implemented with diode-connected or Zero-Vgs TFTs (M1 and M5 in Fig. 6.14a and b respectively). The first solution, although very fast, limits the gain to about 2 and keeps the output common mode very close to ground, making almost impossible the use of additional level shifters. In the second case, the speed worsens, but higher gain can be achieved (up to 40 dB with some design effort) and the output common mode voltage can be designed to be closer to half the supply. For these reasons, Zero-Vgs loads have been preferred in this work.

A third solution was also proposed in literature: the AC-coupled configuration [5]. However, this requires a considerable increase in circuit complexity and the gain improvement is, in our opinion, not so substantial to justify the additional potential yield loss.

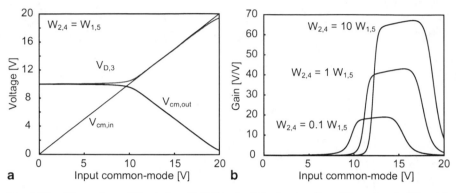

Fig. 6.15 **a** Bias values of the drain of M3 ($V_{D,3}$), of the input voltage ($V_{CM,IN}$) and of the output voltage ($V_{CM,OUT}$). **b** Differential gain as a function of the input common-mode for different dimensioning of the input TFT width. Reprinted from D. Raiteri et al. [4], © 2014, with permission from Elsevier

Amplifier Design and Simulation

The first step in the design of the amplifier shown in Fig. 6.14b is the choice of proper current bias, i.e. the dimensioning of the tail device M3 with respect to the load TFTs, M1 and M5. As starting point, the current variation due to channel length modulation effects can be neglected, and the DC current in the amplifier can be considered independent of the drain-source voltage drop on the loads and on the tail. Since the three devices work in a Zero-Vgs configuration, the tail M3 should provide twice the current provided by M1 and M5. For this reason, the aspect ratio of M3 is chosen $W_3/L_3 = 2\ W_{1,5}/L_{1,5}$. If the input common-mode approaches ground, and both the input transistors are strongly on, the output common mode will approach $V_{DD}/2$.

When the input common-mode approaches V_{DD}, the tail current drastically decreases since the drain-source voltage gets close to zero, i.e. $V_{D,3}$ close to V_{DD} (Fig. 6.15a). However, the DC current will never completely switch off, since the source voltage of the input devices can be even lower than the input common-mode (i.e. $V_{SD,3} > 0$ V even if $V_{CM,\ IN} = V_{DD}$) and the drain voltage of M3 ($V_{D,3}$) never reaches V_{DD}.

The variation of the bias current due to a different input common-mode affects also the output common-mode and the gain of the amplifier. As shown in Fig. 6.15a, when the input common-mode $V_{CM,\ IN}$ is low, both the drain of M3 and the outputs are around $V_{DD}/2$, as expected. When the input common-mode $V_{CM,\ IN}$ becomes larger than half the supply, $V_{D,3}$ follows the inputs and the output common-mode drops, together with the DC current.

From Fig. 6.15b, the consequences on the gain can be observed. For a low input common-mode, the input transistors work in the linear region ($V_{DS} \sim 0$ V and $V_{GS} \gg 0$ V) and the output resistance of the amplifier drops. On the other hand, when $V_{CM,\ IN}$ is too high, the bias current drops, the load transistors M1 and M5 exit

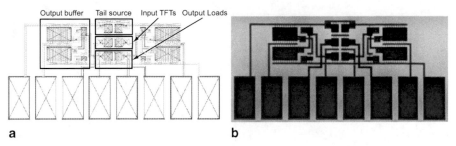

Fig. 6.16 Layout (**a**) and photograph (**b**) of the fully-differential amplifier. Reprinted from D. Raiteri et al. [4], © 2014, with permission from Elsevier

the saturation region, and provide a low output resistance. In both cases, a detrimental effect on the gain is obtained.

Varying the dimension of the input devices causes a different overdrive voltage on the input TFTs (Fig. 6.15b), and hence it shifts the gain plateau to the right for larger widths. Increasing the width of the input devices also increases the gain due to the larger transconductance g_m. Unfortunately, the difference between input and output common-mode increases too, making the design of level shifters for multi-stage DC coupled amplifiers more complicate. Since lowering the width of the input devices also lowers the differential gain (which is already small), the final differential amplifier was designed with input device as wide as the load ones.

None of these considerations are affected by the presence of the second gate. However, some interesting observations can be made for double-gate technologies. Connecting together gate and top-gate of M2 and M4 increases the input transconductance by a factor $1 + \eta$ (in line with Eq. (6.10)), while, concerning the devices used in a Zero-Vgs configuration (M1, M3 and M5), the top-gates should be connected to their own source terminal, in order to prevent detrimental effects on the output resistance due to the threshold voltage variation.

The top-gate voltage could also be used for more sophisticated purposes. For instance, the tail top-gate could be used within a common-mode feedback network to control the output common-mode, and the load top-gates could be used e.g. to zero the offset of the amplifier due to mismatch between M2 and M4, and between M1 and M5.

Amplifier Realization

Figures 6.16a and b show respectively the layout and the photograph of the realized circuit. Both the input TFTs and the output loads have been dimensioned with the same W/L ratio and with the same channel length. Therefore, also the number of sub-channels was chosen the same.

Moreover, the tail source should provide exactly the same current provided by the two load transistors. In order to avoid systematic errors that could worsen the performance of the circuit in addition to process variations, the tail transistor was

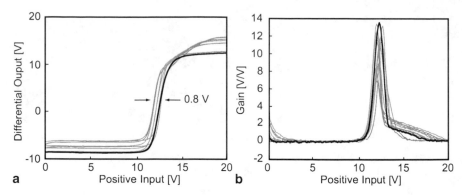

Fig. 6.17 a Measurement of the differential output and **b** maximum gain as a function of the positive input voltage. The negative input was biased at 12 V. Reprinted from D. Raiteri et al. [4], © 2014, with permission from Elsevier

not implemented with a single device exploiting a double width and a different number of sub-channels, but with two devices identical to the loads.

In the final layout, also two output buffers were included. These are required to perform transient measurements without affecting the circuits with external capacitive or resistive parasitics due to the measurement setup.

Measurements

The tapeout includes several instances of the proposed continuous-time amplifier, and in Fig. 6.17 the differential output and gain measured on 14 of them are shown.

For these measurements, the positive input was swept from ground to V_{DD}, while the negative input was kept constant at 12 V. In this way, the differential output was measured and the maximum random offset could be estimated to be about 0.8 V. From these measurements also the maximum differential gain could be evaluated. Indeed, when the positive input reaches 12 V, the input common-mode is also 12 V and, according to the simulations, the maximum gain is achieved. However, the measured gain was much smaller than the simulated one, probably due to the degradation of the semiconductor mobility, and consequently to the reduced input transconductance, after a few weeks shelf-life time. Also its variability is not negligible, showing once more that, even after a careful design process, variations and aging strongly impact analog circuit performance.

6.2.2   A Discrete-Time Amplifier

The performance of continuous-time amplifier suffers from various limitations: low gain, small input range, low speed, low linearity, large mismatch, to mention the most relevant. Many of these drawbacks can be partially solved with a considerable

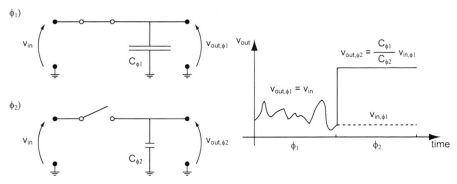

Fig. 6.18 Discrete-time amplification exploiting a parametric capacitor. Reprinted from D. Raiteri et al. [4], © 2014, with permission from Elsevier

increase in the circuit complexity [6], which can impact negatively the yield. Moreover, in our interface, in order to preserve the SNDR achieved by the filter, the amplification should respect constraints on linearity and noise. To achieve these goals, a discrete-time solution has also been investigated.

In feedback-based systems, discrete-time amplification can potentially allow better linearity than continuous-time amplification, as it uses only capacitors as passive devices. Capacitors, contrary to linear resistors, are available in basically any TFT technology on foil, as they can be formed between the two interconnection layers. Unfortunately, the poor performance of large-area technology TFTs makes also the design of discrete-time circuits difficult. High-quality switches cannot be embodied by normally-on transistor without large voltages to switch them off (low off-resistance) and the on-resistance is typically large (causing voltage drops and slow response). Moreover, the most known structures to cancel charge injection, to reduce offsets and leakages, all exploit negative feedback configurations, which are almost impossible to realize as discussed before. For these reasons, a new device was designed to minimize the number of switches required for the sampling, to amplify the signal and to enable faster response.

The Proposed Amplifier

In order to perform fast analog amplification a discrete-time parametric amplifier was designed. The amplifier works in two phases (Fig. 6.18). During the first phase the input signal is connected to a parametric capacitor (PC) and the output voltage $v_{out,\varphi1}$ follows the input v_{in}.

At the beginning of the second phase, the switch opens and the charge accumulated on the capacitance $C_{\varphi1}$ is fixed:

$$Q = C_{\varphi1} \cdot v_{in,\varphi1}. \tag{6.8}$$

where the value $v_{in,\varphi1}$ is the last input voltage before the switch is opened. The second phase also triggers the variation of the capacitance value of the parametric capacitor. For this reason, the charge accumulated in the capacitor in the second phase can be expressed as:

$$Q = C_{\varphi 2} \cdot v_{out,\,\varphi 2}.$$

(6.9)

Imposing charge conservation the gain of the circuit results:

$$G = \frac{v_{out,\,\varphi 2}}{v_{out,\,\varphi 1}} = \frac{C_{\varphi 1}}{C_{\varphi 2}}.$$

(6.10)

Since the two values of the parametric capacitor are independent of the applied voltage, the gain is always signal independent providing intrinsic linearity. Moreover, compared to switched-capacitor discrete-time solutions, the parametric amplifier requires fewer devices (no OpAmps are needed) and needs no charge transfer in the gain phase, with an inherent benefit with respect to robustness and speed.

A Proposed Parametric Capacitor

In order to implement the parametric capacitor, a new device was designed specifically oriented to three-metal-layer technologies. The capacitance value C can be expressed as function of its geometric parameters using the well-know parallel plate formula:

$$C = \varepsilon_0 \varepsilon_r \frac{WL}{h}$$

(6.11)

where ε_0 and ε_r are the electric and the dielectric constant respectively, W and L are the width and the length of the two facing plates, and h is their distance. In a double-gate technology, normal capacitors can be realized between gate and source layers, between gate and top-gate layers, and between the source and top-gate layers. Each couple of layers is characterized by a different value of the parameter h.

Combining in a single device the three types of capacitor, a novel parametric capacitor can be designed. Its cross-section is shown in Fig. 6.19. In this device

Fig. 6.19 Vertical section of the parametric capacitor (PC). The actual width of the second conductive layer depends on the accumulation status in the semiconductor (*green layer*) and varies from the source finger width (FW) to the gate width (W$_G$). (© 2013 IEEE. Reprinted, with permission, from D. Raiteri et al. [2])

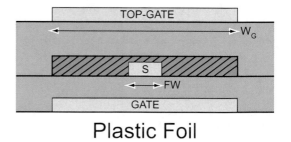

Fig. 6.20 $V_{TG}-V_G$ plane with charge accumulation regions defined by Eq. (6.12). (© 2013 IEEE. Reprinted, with permission, from D. Raiteri et al. [2])

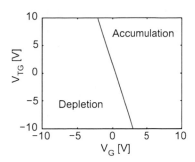

the dimensions of top-gate and gate plates are defined by the geometry of the metal armatures; the middle plate, on the other hand, has a parametric width. In our application, a synchronous signal will be applied to the device in order to accumulate or to deplete the semiconductor. These two states correspond to different dimensions to the central plate, hence the three capacitors can be switched from their minimum to their maximum capacitance value and vice versa. This variation will be used afterwards to amplify the sampled signal.

For our purpose, the top-gate voltage is used to control the accumulation state in the semiconductor underneath. Assuming the S terminal to be ground, and combining Eq. (4.10) with the channel accumulation condition in presence of the top-gate bias $V_G > V_T$, two different regions can be found, in the $V_{TG}-V_G$ plane (Fig. 6.20): on the right hand side of the straight line

$$V_{TG} = -\frac{V_G}{\eta} + \frac{V_{FB}}{\eta} \tag{6.12}$$

the semiconductor is accumulated, while on the left hand side it is depleted.

The new device can be modeled (Fig. 6.21) with three variable capacitors connecting all plate pairs formed by the three metal layers: a top-gate to source (C_{TGS}), a top-gate to gate (C_{TGG}) and a source to gate (C_{SG}) capacitor.

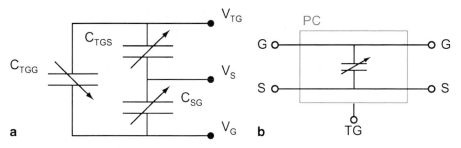

Fig. 6.21 **a** Equivalent model of the parametric capacitor. Three variable capacitors connect top-gate to source (C_{TGS}), top-gate to gate (C_{TGG}) and source to gate (C_{SG}). An increase of C_{TGS} and C_{SG} is associated with a decrease of the capacitance C_{TGG}. **b** Symbol of the parametric capacitor. (© 2013 IEEE. Reprinted, with permission, from D. Raiteri et al. [2])

Based on geometrical considerations it is possible to define for each capacitor the value of its capacitance when the semiconductor is accumulated and when it is depleted. In the case of a depleted channel we have:

$$
\begin{cases}
C_{TGS} = FWL_G \eta C_{ox} \\
C_{TGG} = (W_G - FW)L_G \dfrac{\eta}{\eta+1}C_{ox} \, . \\
C_{SG} = FWL_G C_{ox}
\end{cases}
\tag{6.13}
$$

On the other hand, for an accumulated channel, we find:

$$
\begin{cases}
C_{TGS} = W_G L_G \eta C_{ox} \\
C_{TGG} = 0 \\
C_{SG} = W_G L_G C_{ox}
\end{cases}
\, .
\tag{6.14}
$$

In the previous equations, W_G is the width of gate (G), top-gate (TG) and semiconductor (OSC) layers, FW is the width of the source metal strip (Fig. 6.19) and L_G is the length of the device (in the plane perpendicular to the picture of Fig. 6.19). The PC design of Fig. 6.19 does not take into account the variations due to the lithographic process and to mask misalignments. However, these problems can be easily overcome slightly changing the design of the device accordingly to the specific use. For instance, in our case, the top-gate was used to control the charge accumulation in the semiconductor (and thus the width of the central plate). Therefore a pyramidal layout can be used, designing $W_G = W_{OSC} + E_{OSC}$ and $W_{OSC} = W_{TG} + E_{TG}$, where W_X is the width of the layer X (G, TG or OSC) and E is the minimum enclosure imposed by the design rules. Following this approach, the plate width W_G in Eqs. (6.13) and (6.14) has to be replaced by W_{TG}, but the ratio between the capacitance values in the two different phases remains the same, and the following discussion remains valid.

Each pair of voltages $V_G - V_{TG}$ determines a capacitance value for each of the capacitors in the schematic of Fig. 6.21a. These values are plotted in a three dimensional space in Fig. 6.22 ($W_G = 50$ μm, $L_G = 50$ μm, FW $= 5$ μm, $C_{ox} = 9.5e^{-17}$ F/μm^2), on the left hand side. This picture shows that far enough from the roll off, the value of the capacitance is almost independent of the voltage applied to the plates. The right hand side of the same figure shows the value of each capacitor as a function of only one voltage, while the others are biased at 0 V. The more gradual transition as a function of V_{TG} reflects the weaker coupling due to the higher distance of the top-gate from the semiconductor.

A behavioral model of the PC has been implemented in Verilog-a to enable analog simulation. As well as for the TFT models, the implemented PC instance can parse the width and length of the device from the circuit schematic. In the model, this information is used to tune the capacitor values (C_{TGS}, C_{TGG}, C_{SG}) as a function of the voltage applied to the device.

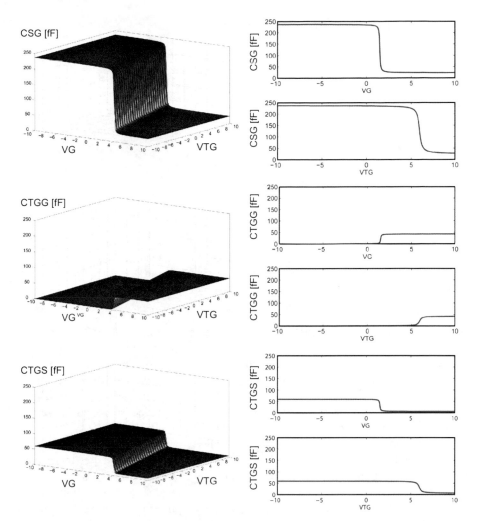

Fig. 6.22 Capacitance values for CSG, CTGG and CTGS: on the left as a function of both gate and top-gate voltages, on the right separately as a function of the gate voltage for $V_{TG}=0$ V (*top panels*), and as a function of the top-gate voltage for $V_G=0$ V (*bottom panels*). (© 2013 IEEE. Reprinted, with permission, from D. Raiteri et al. [2])

Parametric Amplifier Analysis

Based on the former considerations, the PC device can be used to design a discrete-time parametric amplifier. The circuit does not involve active elements and thus does not introduce additional noise, apart from the thermal noise due to the switches involved in the sampling of the analog signal. Moreover, the gain is insensitive to the absolute value of the parametric capacitance, which can be designed to address other requirements like speed ($\tau = C_{PC}R_{switch}$) or noise level (kT/C_{PC}).

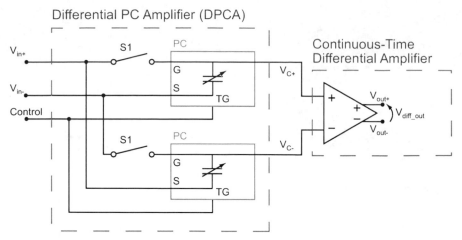

Fig. 6.23 Schematic of the discrete-time amplifier exploiting a differential parametric capacitor amplifier and a continuous-time differential amplifier. Reprinted from D. Raiteri et al. [4], © 2014, with permission from Elsevier

The charge accumulated in the first phase does not need to be transferred to a different capacitor, thus it does not require additional settling time for transport. On the other hand, some time is required to commute the semiconductor state. However, the semiconductor works in accumulation and not in inversion, hence the charge carriers do not need to be borrowed by the contacts as in standard silicon technologies, but just need to be collected at the semiconductor-dielectric interface. For this reason, it is possible to create gain without detrimental effects on speed.

An important point has to be considered: the switching activity of the top-gate metal layer of the new device causes important charge injections on the other plates. The use of two parametric capacitors in a differential amplifier configuration is thus needed to cancel out the spikes due to the charge injection, as the disturbance appears then as a common mode for the differential amplifier that follows them. In the embodiment of Fig. 6.23, a continuous-time differential amplifier, as the one presented in Sect. 6.2.1, is used to process only the differential signal at the output of the parametric capacitor amplifier and drive the following stages of signal conditioning chain, or to provide a suitably large input directly to the analog to digital converter.

In the differential parametric capacitor amplifier DPCA (Fig. 6.23), during the first phase the switches S1 connect the differential input $V_{diff\_in} = V_{in+} - V_{in-}$ to the two PCs. For one of them V_{in+} is connected to the gate and V_{in-} to the source, while for the second PC the gate is connected to V_{in-} and the source to V_{in+}. After the capacitors are charged, the switches S1 are disconnected and the value of the capacitance can be changed varying the control voltage applied to the top-gate. The differential voltage $V_C = V_{C+} - V_{C-}$ is immune from the injection due to the switching control voltage and can be further amplified by the following continuous-time differential amplifier.

Before evaluating the gain provided by the differential PC amplifier, it is important to discuss the missing switch between the differential input and the source terminal of the PCs. Indeed this configuration allows, in the second phase, a direct path from the time variant input to the amplified sampled output. However, the small-signal input applied to the source plate affects the gate plate voltage divided by a capacitive partition (between C_{SG} and C_{TGG}, see Fig. 6.21a) and it is not amplified by the PC; for this reason, its contribution to the differential output should be negligible with respect to the sampled value. If it is not the case, an additional switch exploiting the same function of S1 can be also included, though the new switch would require a different synchronization from S1 and the control signal. In fact, if the switches were opened together, all devices would be floating and all potentials would shift with an equal amount of the control voltage, leaving the state of the semiconductor unchanged. For this reason, the second switch would require a delayed trigger providing enough skew to let the semiconductor commute its state.

Considering the low quality of the TFTs on foil as switch, the reliability issues for complex circuits, and the small accuracy of data converters (that will be discussed in more detail in next chapter), the influence of the direct path of the input to the output was evaluated a second order effect and a solution without additional switches was preferred. Based on this choice, the gain of the differential PC amplifier can be calculated evaluating the output voltage V_C in the second phase applying charge conservation at the output node of the variable capacitors, i.e. the gate plate. Naming V_{low} and V_{high} the two possible levels of the control voltage applied to the top-gate plate, and referring to Fig. 6.21a and Fig. 6.23 for the capacitance names and for the voltages in the discrete-time amplifier, first the behavior of V_{C+} can be evaluated between the two phases. The charge on the gate plate of the device connected to the positive terminal of the continuous-time amplifier reads:

$$Q_1 = (V_{in+} - V_{low})C_{TGG,\varphi1} + (V_{in+} - V_{in-})C_{SG,\varphi1} \tag{6.15}$$

$$Q_2 = (V_{C+} - V_{high})C_{TGG,\varphi2} + (V_{C+} - V_{in-})C_{SG,\varphi2}. \tag{6.16}$$

Imposing the charge conservation, i.e. $Q_1 = Q_2$, V_{C+} can be expressed as:

$$V_{C+} = V_{in+}\left(\frac{C_{TGG,\varphi1} + C_{SG,\varphi1}}{C_{TGG,\varphi2} + C_{SG,\varphi2}}\right) - V_{in-}\left(\frac{C_{SG,\varphi1} - C_{SG,\varphi2}}{C_{TGG,\varphi2} + C_{SG,\varphi2}}\right) + \frac{V_{high}C_{TGG,\varphi2} - V_{low}C_{TGG,\varphi1}}{C_{TGG,\varphi2} + C_{SG,\varphi2}}. \tag{6.17}$$

The first two terms on the right hand side of this expression represent the effect of the differential input, while the last one gives the common-mode contribution due to the switching activity at the top-gate plate.

At the negative input of the continuous time amplifier, the same process takes place, but the inputs are inverted. Therefore:

$$V_{C-} = V_{in-}\left(\frac{C_{TGG,\varphi1} + C_{SG,\varphi1}}{C_{TGG,\varphi2} + C_{SG,\varphi2}}\right) - V_{in+}\left(\frac{C_{SG,\varphi1} - C_{SG,\varphi2}}{C_{TGG,\varphi2} + C_{SG,\varphi2}}\right) + \frac{V_{high}C_{TGG,\varphi2} - V_{low}C_{TGG,\varphi1}}{C_{TGG,\varphi2} + C_{SG,\varphi2}}. \tag{6.18}$$

The amplified signal $V_C = V_{C+} - V_{C-}$ as a function of the differential input V_{in} can thus be written as:

$$V_C = V_{in} \left(\frac{C_{TGG,\varphi 1} + C_{SG,\varphi 1}}{C_{TGG,\varphi 2} + C_{SG,\varphi 2}} + \frac{C_{SG,\varphi 1} - C_{SG,\varphi 2}}{C_{TGG,\varphi 2} + C_{SG,\varphi 2}} \right). \tag{6.19}$$

The differential gain of the differential parametric capacitor amplifier G_{DPCA} results:

$$G_{DPCA} = \frac{C_{TGG,\varphi 1} + 2C_{SG,\varphi 1} - C_{SG,\varphi 2}}{C_{TGG,\varphi 2} + C_{SG,\varphi 2}}. \tag{6.20}$$

The subscript '1' indicates the value of the capacitance in the sampling phase, which takes place with accumulated semiconductor (Eq. (6.13)), and the subscript '2' indicates the value of the capacitance in the second phase, when the semiconductor is depleted (Eq. (6.14)).

Amplifier Design and Simulation

The expression of the discrete-time gain G_{DPCA} can be rewritten, remembering Eqs. (6.13) and (6.14), as a function of the geometry of the device:

$$G_{DPCA} = \frac{2 \dfrac{W_G}{FW}}{\dfrac{W_G}{FW} + \dfrac{1}{\eta}} \frac{\eta + 1}{\eta}. \tag{6.21}$$

This equation analytically shows that the gain is independent of the absolute value of the capacitance; indeed, the length of the device L_G is cancelled out in the ratio between numerator and denominator. In our technology, the coupling of the top-gate is $\eta = 0.25$, which leads, if $W_G \gg FW$, to a maximum theoretical gain equal to

$$G_{DPCA,max} = 2 \frac{\eta + 1}{\eta} = 10. \tag{6.22}$$

The gain of the DPCA increases proportionally the gain of the continuous-time differential amplifier (Fig. 6.23) without any detrimental effect on its speed. Indeed, the bandwidth of the parametric capacitor is much higher than any normal continuous-time differential amplifier topology known in this technology. The increased gain and preserved speed are confirmed in Fig. 6.24, where two simulated transients are shown. The continuous line corresponds to the system exploiting the differential parametric capacitor amplifier together with the continuous-time differential amplifier, while the dashed line is obtained using a just normal capacitor instead of the parametric capacitor. The differential input was set to 10 mV and the simulated gain of the continuous-time differential amplifier $G_{DiffAmpl}$ is about 11.

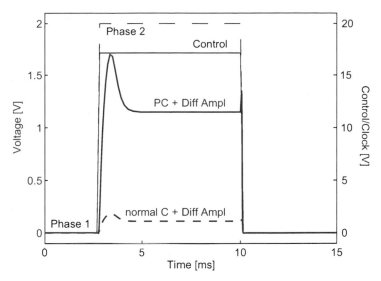

Fig. 6.24 Transient simulation of the proposed parametric capacitor-based frontend compared to one exploiting a normal S&H when a 10 mV differential input is applied. The output signals of the two frontends (*blue lines*) refer to the y-axis on the left, while the control signal and the clock (*black lines*) refer to the y-axis on the right. (© 2013 IEEE. Reprinted, with permission, from D. Raiteri et al. [2])

The shape of the step response is the same in both cases, but the amplitude of the output signal is almost ten times larger using the differential parametric capacitor amplifier, as one would expect from Eq. (6.22).

6.3   Conclusions

In this chapter, some building blocks suitable for analog signal conditioning have been presented. The proposed use of these blocks, in an actual interface for smart sensor applications, exploits an analog filter first and a discrete-time amplifier afterwards. In this way, the linearity requirements on the filter are relaxed, while the amplifier also implements sampling functionality useful for the following analog to digital converter without suffering from aliasing.

The proposed low-pass filter is a G_mC filter, where an innovative transconductor design has been proposed. In order to cope with the considerable variability affecting TFTs manufactured at low temperature, the double-gate feature of our technology was used to provide the circuit with tunability and increased robustness. The transconductance and output resistance tunability was measured to be larger than one order of magnitude. This flexibility can also be used, in more uniform processes, to realize filters with electrically controlled bandwidth. Moreover, the unloaded gain of the circuit is independent of the transconductance, since the output resistance scales accordingly. The tuning effect is thus a rigid shift of the characteristic along the frequency axis.

The filtered signal needs to be amplified to achieve a full scale range compatible with the input range of the analog-to-digital converter. Unfortunately, linear amplifiers are difficult to implement in large-area technologies exploiting unipolar TFTs, and a continuous-time amplifier can only reach a gain of a few tens with a very small common-mode input range. To overcome the performance limits of TFTs, a discrete-time parametric amplifier was also proposed. This amplifier exploits a new parametric capacitor, which was designed specifically for our technology, but can be applied to any three-metal-layer thin-film process. The parametric amplifier has intrinsic linearity and allows for a larger input common-mode since the amplification is independent of the DC input and output voltages. Also with respect to process variations, the proposed device should perform better than conventional differential amplifiers, since the amplification only depends on the thickness of the insulator layers and not on masks alignment or lateral dimensions, neither on device parameters like semiconductor mobility and threshold voltage.

References

1. D. Raiteri, F. Torricelli, E. Cantatore, A.H.M.V. Roermund, A tunable transconductor for analog amplification and filtering based on double-gate organic TFTs. *IEEE European solid-state circuits conference*, pp. 415–418, 2011
2. D. Raiteri, A.H.M.V. Roermund, E. Cantatore, A discrete-time amplifier based on thin-film trans-capacitors for organic sensor frontends. *IEEE international workshop on advances in sensors and interfaces*, 2013
3. D. Raiteri, F. Torricelli, P.V. Lieshout, A.H.M.V. Roermund, E. Cantatore, A synchronous rail-to-rail latched comparator based on double-gate organic thin-film-transistors. *IEEE European solid-state circuits conference*, pp. 141–144, 2012
4. D. Raiteri, A.H.M.V. Roermund, E. Cantatore, A discrete-time amplifier based on thin-film trans-capacitors for sensor systems on foil. Microelectron. J. (2014). doi: 10.1016/j.mejo.2014.09.002
5. H. Marien, M. Steyaert, N.V. Aerle, P. Heremans, A mixed-signal organic 1 kHz comparator with low VT sensitivity on flexible plastic substrate. *IEEE European solid-state circuit conference*, pp. 120–123, 2009
6. H. Marien, M.S.J. Steyaert, E. van Veenendaal, P. Heremans, A fully integrated $\Delta\Sigma$ ADC in organic thin-film transistor technology on flexible plastic foil. IEEE J. Solid-State Circuits **46**(1), 276–284, 2011

Chapter 7
Circuit Design for Data Conversion

After the physical signal is converted to the electric domain, filtered and amplified, it must be quantized and converted into a digital word. The analog to digital
conversion can be performed following many different approaches; the chosen one
should take into account various aspects, like application specifications and signal
properties. In our case, the main constraints come from the technology of TFTs on
foil. In this chapter, circuit implementation for some building blocks commonly
used for data conversion according to the scheme of Fig. 5.2 will be proposed: first
a synchronous comparator and then a current-steering digital-to-analog converter.
In the last part of this chapter, a complete integrating analog-to-digital converter is
analyzed, designed and characterized. Parts of this chapter have been published in
[1–3].

7.1 A Synchronous Latched Comparator

Comparators are among the most common and most critical building blocks in ADC
architectures (Fig. 5.2). Indeed, in a careful converter design, they play a significant role in determining the noise level, conversion speed and, depending on the
architecture, linearity of the ADC. A very explanatory example is the basic flash
architecture, which requires just one clock cycle to convert the analog signal in a
thermometric digital word, and whose accuracy and linearity is limited by the input
noise and the offset statistics of the 2^{n-1} comparators (and by the statistical variation
of the reference voltages).

7.1.1 Comparator Topology and Analysis

A common solution [4] to implement a fast comparator, able to resolve small
differential input signals and achieve logic output levels, is shown in Fig. 7.1.

© Springer International Publishing Switzerland 2015   81
D. Raiteri et al., *Circuit Design on Plastic Foils,* Analog Circuits and Signal Processing,
DOI 10.1007/978-3-319-11427-9_7

Fig. 7.1 Comparator exploiting a preamplifier with
output offset cancellation
and synchronous latch for
logic levels generation. (©
2012 IEEE. Reprinted, with
permission, from D. Raiteri
et al. [1])

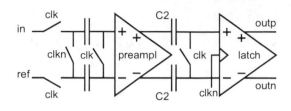

The logic levels are regenerated by the latch that, exploiting a positive feedback, forces the two outputs to saturate to different supply rails, according to the differential input. When designing a comparator for high speed, latches must have small input devices to reduce parasitic capacitance, but, for this reason, their input offset (i.e. the input voltage required to compensate for all causes of mismatch and obtain zero output voltage) is typically large (in IC silicon technologies the mismatch between the input devices, which typically is the dominant cause of offset, has a standard variation inversely proportional to the square root of the channel area [5]). In order to improve the accuracy of the comparator, a preamplifier (Fig. 7.1) can be employed to reduce the input equivalent offset due to the latch with a factor equal to the gain of the preamplifier $G_{preampl}$. The amplifier can be optimized for speed too, i.e. choosing low gain (below 10) and high bandwidth, since its offset can be easily cancelled out after sampling it at the preamplifier output, on the capacitances C2 in Fig. 7.1, using the scheme that is typically called output offset cancellation.

Following this approach, the continuous-time fully differential amplifier shown in the previous chapter could be used, while the design of a synchronous latch taking advantage of the double-gate technology is shown in the rest of this section.

A basic way to implement a latch is to connect two transconductors G_m in a positive-feedback configuration (Fig. 7.2a). In our technology, a transconductor can be implemented employing a Zero-Vgs connected TFT as active load and a common source TFT as a driver. Figure 7.2b shows the transistor level implementation of

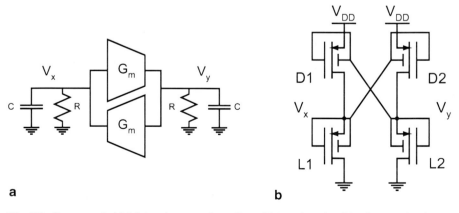

a   b

Fig. 7.2 Cross-coupled latch topology: **a** schematic and **b** transistor level implementation in our double-gate technology. (© 2012 IEEE. Reprinted, with permission, from D. Raiteri et al. [1])

the cross-coupled G_m, where the gate and the top-gate terminals of the driver are connected together to improve the transconductance by a factor $1+\eta$, according to Eq. (4.10) [6]. In this implementation, the transconductance G_m in Fig. 7.2a is equal to the transconductance g_m of the drivers D1 and D2 in Fig. 7.2b. The resistance R in Fig. 7.2a represents the output resistance of the Zero-Vgs loads L1 and L2 in parallel with the output resistance of the drivers, and the capacitance C takes into account the overall capacitance loading the output nodes V_x and V_y.

These three parameters (G_m, R and C) define the time behavior of the circuit, which can be evaluated solving the differential equations at nodes V_x and V_y. Indeed, when a small imbalance ΔV_0 is applied between the output nodes x and y, the positive feedback regenerates the imbalance, eventually bringing the outputs to the opposite rails. In order to derive analytically the trend of the initial imbalance, first the behavior of V_x and V_y mast be written as

$$V_x = -\frac{C}{G_m}\frac{\delta V_y}{\delta t} - \frac{V_y}{G_m R}, \tag{7.1}$$

$$V_y = -\frac{C}{G_m}\frac{\delta V_x}{\delta t} - \frac{V_x}{G_m R}. \tag{7.2}$$

Subtracting Eq. (7.2) from Eq. (7.1), a differential equation in $\Delta V = V_x - V_y$ is obtained:

$$\Delta V = \frac{C}{G_m}\frac{\delta \Delta V}{\delta t} + \frac{\Delta V}{G_m R}. \tag{7.3}$$

Then, the transient behavior of ΔV can be written as:

$$\Delta V = \Delta V_0 e^{t/\tau}, \tag{7.4}$$

where the time constant τ is expressed as:

$$\tau = \frac{RC}{G_m R - 1}. \tag{7.5}$$

In order to take real advantage of the positive feedback the gain $G_m R$ of the single transconductor needs to be much larger than 1. Under this assumption, Eq. (7.5) can be simplified as:

$$\tau \sim \frac{C}{G_m}. \tag{7.6}$$

This formulation of the time constant associated to the circuit highlights that the speed achieved by this circuit is only limited by the technology. Indeed, it depends on technology features like the parasitic capacitance of the transistors and on their transconductance, which is proportional to the semiconductor mobility.

The time t_f for the output to reach a certain differential voltage ΔV_f, however, does not only depend on the time constant τ, but also on the value of the initial imbalance ΔV_0, and can be calculated from Eq. (7.4):

$$t_f = \tau \log \frac{\Delta V_f}{\Delta V_0}. \tag{7.7}$$

The variation of the comparison time Δt_f due to the variation of the initial unbalance ΔV_0 will be referred to in the following as time-walk. The comparison is slower when a small differential input is fed to the circuit, and faster when the differential input is larger.

7.1.2   The Proposed Latched Comparator

In a synchronous implementation of the latched comparator, the cross-coupled inverter latch requires an additional transistor MC, connected to the clock reference CLK, to synchronize the comparison, and two capacitors C_{dec} to provide the first imbalance and allow the output nodes to diverge (Fig. 7.3).

Unfortunately, the Zero-Vgs load charges the output with a large parasitic capacitance C_p (due to its gate and top-gate); in fact, in order to reach correct output logic levels, the load must be designed with a W/L much wider than that of the driver (in this case a factor 12.5 provides good results in simulation). The large parasitic capacitance on the output nodes is a major drawback of the topology shown in Fig. 7.3. The capacitors C_{dec} indeed isolate the output nodes from the input source and apply the input differential voltage to the latch. The input signal is thus divided between the total output capacitance of the latch (shown in Fig. 7.3 with $C_p \sim 2.6\text{pF}$)

Fig. 7.3 Schematic of the cross-coupled inverter based latch. (© 2012 IEEE. Reprinted, with permission, from D. Raiteri et al. [1])

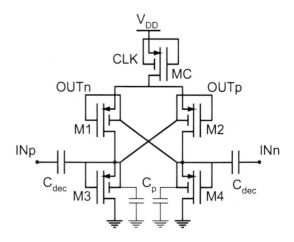

and the decoupling capacitance C_{dec}, reducing the initial imbalance ΔV_0. In order to avoid a significant loss of signal, the decoupling capacitances have to be much larger than C_p. If the $C_{dec} = DC_p$, the input imbalance becomes about $\Delta V_0 (1 - 1/D)$ and the actual time constant τ' of the circuit can be expressed as:

$$\tau' \sim D\tau = \frac{C_p + C_{dec}}{C_p} \frac{C_p}{G_m}. \tag{7.8}$$

This equation shows that if a voltage division of a factor $(1 - 1/D)$ can be allowed at the input, an increase of the time constant time by a factor D has to be accepted. Another issue related to this topology is that the input signal has to be applied almost together with the active clock transition to avoid the discharge of C_p through the Zero-Vgs load transistors, which are always slightly on.

To solve this problem, advantage can be taken of the double-gate feature of our technology and design a new latch where inputs and outputs are decoupled without the need for explicit capacitors. As depicted in Fig. 7.4, in this solution the input is applied directly to the driver of the inverters through the top gates of M5 and M6. This avoids the use of the large decoupling capacitances C_{dec} and brings a considerable gain in area, yield and speed.

In the circuit of Fig. 7.3, the input capacitance was large due to the parasitics at the gate and at the top-gate of the wide load transistor. In the circuit of Fig. 7.4, where the signal is fed to the top-gate of the much smaller driver device, the input capacitance of the circuit is strongly reduced. Thanks to this solution, the input capacitance can be made about two orders of magnitude smaller, and also the circuitry driving the comparator will need to source a correspondingly lower current. Furthermore the differential voltage applied to the top-gate of the input pair M5-M6 creates an imbalance in the threshold voltage on a fully isolated node, thus this comparator does not require a synchronous input, but can perfectly work also with quasi-static input signals.

The drawback of the configuration in Fig. 7.4 is that it reduces the transconductance of the input devices. Therefore, the positive feedback becomes weaker and the settling time is increased, since the transconductance in Eq. (7.6) is lower. Moreover

Fig. 7.4 Schematic of the input-decoupled latch. (© 2012 IEEE. Reprinted, with permission, from D. Raiteri et al. [1])

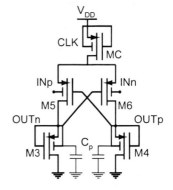

the output logic level "high" of the comparator becomes dependent on the input common mode and worsens due to the weaker pull up. This can result in the need for an additional stage to regenerate full-swing digital levels.

The two topologies shown so far present both some interesting features, such as high gain and good output logic levels in the cross-coupled inverters (Fig. 7.3), and high speed and low input capacitance in the input-decoupled one (Fig. 7.4). However, the first is slow and the second less robust.

By merging the two previous topologies, the final latched comparator shown in Fig. 7.5 achieves higher global performance. In fact, it adopts both a driver pair with gate connected to the top-gate (M1-M2) and top-gate input transistors M5-M6. By means of this merging, the latched comparator guarantees enough gain to reach a full swing of the outputs still without compromising the speed of the response.

7.1.3   Latch Design and Simulation

In order to prove the actual speed improvement of the proposed latch (Fig. 7.5) with respect to the basic implementation (Fig. 7.3), the transient simulations of both circuits are shown in Fig. 7.6.

When the clock signal commutes from V_{DD} to ground, the two output signals grow and the positive feedback starts splitting the outputs accordingly to the first imbalance. The proposed latch (Fig. 7.5) reaches the same logic levels as the basic one (Fig. 7.3), providing robust propagation of the comparison result, and it can complete both comparison and output reset in less than 5 ms (Fig. 7.6a), compared to about 35 ms (Fig. 7.6b) required by the basic latch topology. This difference is mainly due to the presence, in the basic latch (Fig. 7.3), of the decoupling capacitances which allow the output nodes to regenerate logic levels independently from the constant differential input. In the new topology, since the signal is fed through the top-gates of M5 and M6, the decoupling capacitors are not required.

Fig. 7.5 Schematic of the synchronous rail-to-rail latch. (© 2012 IEEE. Reprinted, with permission, from D. Raiteri et al. [1])

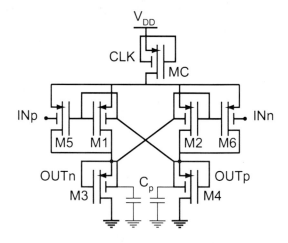

Fig. 7.6 Transient simulation of the output signal of the **a** proposed latch (Fig. 7.5) and of the **b** basic latch (Fig. 7.3). The speed of the response depends also on the first imbalance entity: the larger the initial imbalance, the faster the response

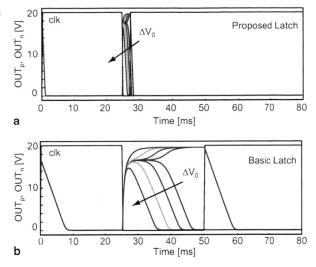

Thus, the response is about ten times faster even if the ratio between drivers (M1, M2, M5 and M6) and loads (M3 and M4) is the same. Indeed, for the proposed comparators, the loads have the same dimensions as for the basic one, i.e. $W_{3,4}/L_{3,4}=2000$ μm/20 μm, while in the former case $W_{1,2,5,6}/L_{1,2,5,6}=40$ μm/5 μm and in the latter $W_{1,2}/L_{1,2}=80$ μm/5 μm. In other words, in the proposed latch, the original input device is split in two parts, one to generate the positive feedback gain and the other to read the input. Also the reset phase is very important, because a residual voltage on one of the outputs would reflect in signal dependent memory effects. The simulation in Fig. 7.6 was launched for different initial imbalances ($\Delta V_0 = 1$ μV, 10 μV, 100 μV, 1 mV), and the consequent time-walk (Eq. (7.7)) can be estimated.

Figure 7.6a shows, with continuous lines, the simulated time evolution of the differential output voltage of the latch (after the buffers) for a constant common mode $V_{cm} = 18$ V and varying the differential input ΔV_0. The comparison starts when the clock signal switches from high to low, and, in line with Eq. (7.7), a time-walk as a function of the differential input ΔV_0 is observed. Fitting Eq. (7.7) to the simulation results, a time constant $\tau = 210$ μs is obtained, while the delay to reach 50% of the output for $\Delta V_0 = 100$ mV is 1.6 ms.

Measuring a circuit that bases its performance on the large output resistance and the small parasitic capacitance is a delicate task. Indeed, a measurement setup for dynamic characterization typically loads the measurement pad with a relatively low resistance (1 MΩ) and large capacitance (up to 100 pF, due to the long interconnects). For this reason, an inverter and a custom designed buffer [7] were integrated after each output. Even using these circuits the effect of the measurement setup is not negligible and was also inserted in the simulation after the circuit realization to estimate the quality of the dynamic model.

The simulation results in presence of the output buffer and measurement load are shown in Fig. 7.7a, in the lower plot.

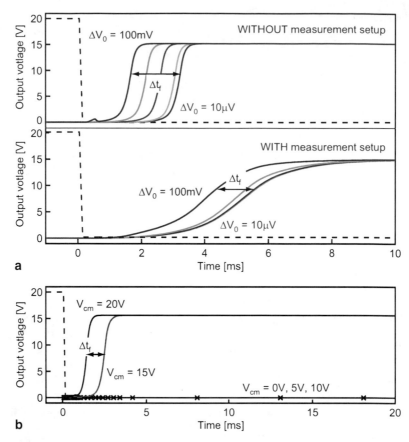

Fig. 7.7 Simulated time-walk (Δt_f) **a** due to varying differential input ($\Delta V_0 = 10\ \mu V$, $100\ \mu V$, $1\ mV$, $10\ mV$, $100\ mV$) and same input common mode ($V_{CM} = 18\ V$), and **b** due to varying input common mode ($V_{CM} = 0\ V$, $5\ V$, $10\ V$, $15\ V$, $20\ V$) and same differential input ($\Delta V_0 = 10\ mV$). *Dashed lines* represent the synchronization signal. (© 2012 IEEE. Reprinted, with permission, from D. Raiteri et al. [1])

An important feature of the proposed latch is the broad common mode input range allowed. Indeed the input common mode influences two (G_m and R) out of three parameters (see Eq. (7.5)) determining the comparison time and, in some circumstances, it can also prevent a successful comparison (if $G_m R < 1$), as shown in Fig. 7.7b for $V_{cm} < 10\ V$.

For a differential input ΔV_0 set at $10\ mV$, the transient of the outputs was simulated for common-mode voltages V_{cm} varying from $0\ V$ to $20\ V$ in steps of $5\ V$. In this case, a faster response is obtained with higher input common mode levels, and for $V_{cm} \leq 10\ V$ the comparison does not take place at all. This behavior is coherent with the effects the top-gate has on the TFTs. Indeed, for low top-gate voltages, the threshold of TFTs becomes so positive that M5 and M6 start working in the linear region. In this case, the positive feedback is too weak (or absent), and the resulting response slow. On the other hand, when the input common mode is high, the two

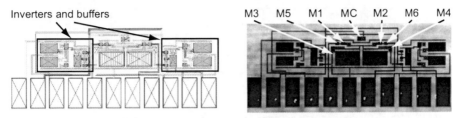

Fig. 7.8 Layout and photograph of the proposed latch. The proposed latch is placed in the middle, while on each side are one inverter and one buffer to drive the measurement setup

input devices work in their saturation region providing the necessary output resistance to make the positive feedback effective.

7.1.4 Latch Realization

The proposed latch was realized and measured. The layout and the photograph are shown in Fig. 7.8.

In the layout, the inverter and the buffer integrated to drive the measurement setup are highlighted, while, in the photograph of the circuit, the position of each transistor is shown.

On the bottom, the pads reveal the fragility of the surface of the plastic foil. Indeed, each pad has a clear sign were the probe landed. Also, the number of pads is larger than the inputs and outputs of the circuit. In fact, additional pads have been included to be able to measure each transistor separately and compare their electrical properties, and to get a feeling of the matching between input devices and between output devices.

7.1.5 Latch Measurements

The corresponding measurements on the actual circuit, after the buffers, are shown in Fig. 7.9. Figure 7.9a shows the time-walk for a constant $V_{cm} = 18$ V and a varying differential input ($\Delta V_0 = 2.5$ V, 1.5 V, 0.5 V, 0.3 V, 0.1 V—values corrected for offset[1]). Figure 7.9b plots the time response for $V_{cm} = 10$ V, 12 V, 14 V, 16 V, 18 V and a constant $\Delta V_0 = 100$ mV. The results show, in agreement with simulations, that for common-mode voltages lower than 10 V the comparator does not work properly. In the two plots, the black lines show the clock signal that synchronizes the comparison.

---

[1] The offset was estimated while measuring the output voltage as a function of the input common mode (Fig. 7.10). It was evaluated as the average between the minimum initial imbalance that always provides a positive comparison result and the maximum initial imbalance that always provides a negative comparison result. If the initial imbalance is between these two, the input noise randomly determines the polarity of the comparator output.

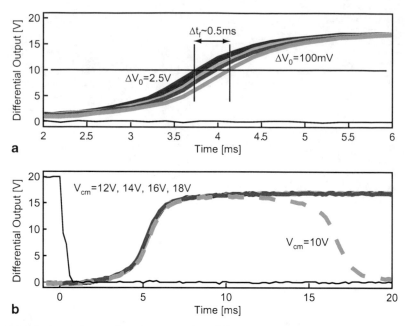

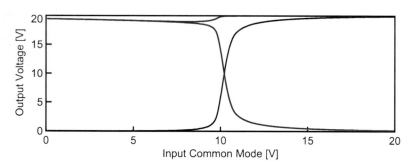

Fig. 7.9 Measured time-walk **a** due to varying differential input and constant input common mode ($V_{CM} = 18$ V), and **b** due to varying input common mode and constant differential input ($\Delta V_0 = 100$ mV). (© 2012 IEEE. Reprinted, with permission, from D. Raiteri et al.[1])

Fig. 7.10 Measured output voltage as a function of the input common mode V_{cm} with a constant $\Delta V_0 = 100$ mV (after subtracting the offset). The *blue*, *red* and *black lines* represent respectively the positive, the negative and the differential output. (© 2012 IEEE. Reprinted, with permission, from D. Raiteri et al. [1])

To get deeper insight in the dependence of the gain on the common mode input voltage, the single-ended and the differential outputs are shown in Fig. 7.10, as a function of the input common mode, applying a static differential input $\Delta V_0 = 100$ mV. The latched comparator, measured using a semiconductor parameter analyzer with extremely high input impedance, can amplify the small differential input to a rail-to-rail differential output, meaning that the small-signal differential gain must be higher than $G_{diff} = 20$ V/100 mV $= 200$ V/V. Such high gain, enabled

by the positive feedback, is very beneficial to reduce the effect of the noise of the
input devices on the comparison. Still from Fig. 7.10, the minimum input common
mode needed for proper operation is estimated around 10 V, in line with what was
concluded based on the CAD simulations and dynamic measurement. It is worth
noting that the differential gain is not the slope of the black line; indeed each point
in Fig. 7.10 represents the value of the outputs (or their difference) after the transient is completed, and the steady state has been reached.

As stated in the beginning of the chapter, an important requirement on the comparators static performance is the input referred offset which in some converter
topologies sets the linearity limit. Since matching simulations could not be run due
to the lack of suitable characterization, some additional test pads were included in
the layout to be able to measure each transistor in the latch individually.

In this way, it was possible to measure and compare (Fig. 7.11a) the currents
flowing through the input devices when diode connected (i.e. with inputs and
outputs to ground) and the relative error of one current with respect to the other
(inset). In a similar way, the output currents of the loads were compared. In this case
however, the currents do not have the shape of a diode current, but of a Zero-Vgs
connected TFT (Fig. 7.11b). The mismatch between the two drivers and between
the two loads causes an offset of the latch of about $V_{off}=3.4$ V. This offset is particularly large, even for large-area low-temperature technologies. One reason can be
found in the weak top-gate transconductance. Indeed, let us consider, for instance,
the flat-band mismatch ΔV_{FB} affecting the input devices. It contributes to the input
offset with $V_{of,\,gate}=\Delta V_{FB}$ if the input is applied to the gate, but it is $1/\eta$ times larger
(i.e. $V_{of,\,top\text{-}gate}=\Delta V_{FB}/\eta$, in our case $1/\eta=4$) if the input is applied to the top-gate.

Exploiting the preamplifier shown in the previous chapter, which average
gain is $G_{preampl}=12$, the comparator can achieve an input referred offset of about
$V_{of,\,input}=280$ mV. Considering an input common-mode range of 10 V, this offset
value makes the comparator suitable for a 5 bit flash ADC with a full scale input
range. A continuous-time amplifier with higher gain would allow lower offset
and higher accuracy, but the speed of the circuit would decrease too. A better
solution would be to interpose between the preamplifier and the latch a differential

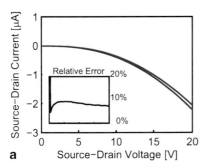
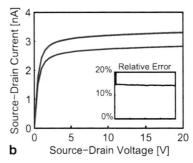

Fig. 7.11 Evaluation of the matching between **a** input and **b** output transistors. The corresponding relative error is shown in their own inset. (© 2012 IEEE. Reprinted, with permission, from D.
Raiteri et al. [1])

parametric-capacitor amplifier (DPCA) like the one shown in Sect. 6.2.2. Indeed, the common-mode signal generated by the switching activity of the DPCA would be cancelled out by the differential input of the latch, and the low input capacitance of the latch would influence the output of the DPCA even less than the classic differential pair.

The latch dissipates a static power of about 30 nW when holding the result of the comparison. Due to the switching activity, a dynamic power depending on the capacitance loading at the outputs of the latch is also dissipated. For instance, for $f_{clk} = 100$ Hz the dynamic power dissipated is $P_{dyn} \sim 100$ nW.

7.2 A Digital to Analog Converter Based on a Metal-Oxide Technology

At the end of the previous section, the linearity achievable by AD converters exploiting a flash topology was related to the statistics of the input offset of the latch. If the variability is too large, alternative topologies should be considered. For instance, in SAR ADCs, only one comparator is used and the input offset of the comparator only causes a rigid shift of the static characteristic without any detrimental effects on the linearity. On the other hand, linearity is determined by the linearity of the reference signal which is compared to the input signal, and that is generated by a DA converter. C-2C DACs for SAR ADCs have already been explored, leading to a maximum INL of 3 LSB for a resolution level of 6 bit before calibration [8, 9]. In the intent of investigating other DAC solutions, a current-steering DAC was analyzed, designed and characterized as described in this section.

7.2.1 GIZO Technology

For this purpose, a current-steering topology was investigated exploiting a metal-oxide technology that has been recently proposed [10]. This technology employs amorphous Gallium–Indium–Zinc–Oxide (GIZO or IGZO), which is an interesting semiconductor for manufacturing TFTs on foil because its mobility ($\mu \sim 20$ cm^2/Vs) is superior to other common materials for large-area electronics (organic and a-Si have mobility $\mu \sim 1$ cm^2/Vs). The amorphous nature of GIZO grants also a good uniformity, contrary to Low Temperature Polycrystalline Silicon (LTPS), which still offers the best mobility among large-area TFT technologies ($\mu \sim 100$ cm^2/Vs), but at the cost of larger variability. The optical transparency and the relatively low fabrication temperature ($< 150\,^\circ$C) make this technology also especially suitable for display backplanes and related driving electronics [11], as well as for any kind of large-area applications on plastic foils, e.g. biomedical sensors, non-volatile memories [12], RFIDs [13], and alike.

For circuit simulations, a physics-based analytical model of the GIZO TFTs has been developed in house similar to the one proposed in Chap. 4, and taking into

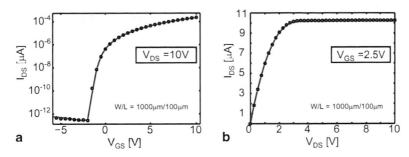

Fig. 7.12 Measured (*symbols*) and modeled (*lines*) **a** transfer and **b** output characteristics of a GIZO TFT (W = 1000 μm, L = 100 μm)

account a Multiple Trapping and Release (MTR) charge transport in the channel of the transistor, instead of Variable Range Hopping (VRH). Also in this case, the Density of States can be considered exponential. The model is fully symmetrical and accurately describes (Fig. 7.12) the below-threshold, the linear and the saturation regimes via a unique formulation.

For transient simulations, overlap capacitances between gate, source and drain have also been included (Fig. 7.13). In our simple model, both gate-source and gate-drain capacitors have a constant overlap component $C_{overlap}$ (much larger than in self-aligned silicon technologies), proportional to the finger width (FW) of the source and drain contacts, and a component depending on the channel accumulation which is divided between the two contacts. The parameterized cell (PCell) developed for our CAD system, only accepts an even number of sub-channels SC, leading thus to a smaller gate-drain capacitance C_{GD} compared to the gate-source C_{GS} one:

$$C_{GS} = C_{GD}\left(1 + \frac{2}{SC}\right). \tag{7.9}$$

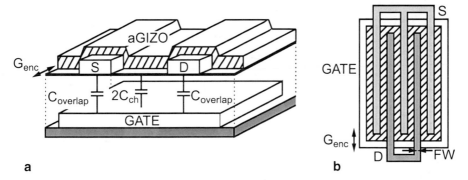

Fig. 7.13 **a** Stack and **b** layout of the implemented GIZO TFTs

Moreover, the source metal is also connected to the periphery of the semiconductor layer that hence contributes only to the capacitance between gate and source.

7.2.2   The Proposed Current-Steering DAC

Besides the possible use in smart sensors, one of the most promising applications for GIZO technologies are display backplanes where a DAC exploiting GIZO TFTs is integrated at the periphery [14, 15] of a display to generate the analog signals that define the image content. The adoption of the same technology for electronics and backplane avoids the use of separate silicon chips and complex interconnects, resulting in large potential cost savings. Crucial to this application are resolutions between 6 bit and 8 bit along with adequate speed (e.g. the row selection time for a video QVGA display is 68.3 µs). GIZO TFTs are suitable to build DACs due to their good mobility and uniformity, but so far only simple amplifiers, and no data converters, have been reported in literature [16, 17].

Switched capacitor, resistive or current-steering (CS) approaches can be adopted to implement a TFT DAC in this context. The first two implementations [14, 15] rely on the relative accuracy of passive components, and are thus suited to the LTPS technology, which offers poor TFT uniformity. On the other hand, the CS approach is particularly interesting in combination with AMOLED displays, as it allows current programming in the pixel [2], which can effectively compensate for both threshold and mobility variations in the backplane: an attractive solution to obtain excellent OLED uniformity and increase resilience to ageing.

Current-Steering DAC Analysis

The current-steering DAC is a converter that produces a current proportional to the digital input word. In our implementation, every bit of the digital input D_{IN} controls a binary weighted current so that, for N bit resolution, it is possible to generate 2^{N-1} current levels. In the case a voltage signal is required at the output, the current can be converted into a voltage by means of a linear resistor, as depicted in Fig. 7.14.

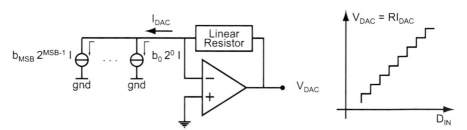

Fig. 7.14 Current-steering DAC topology exploiting a resistor in a negative feedback for linear I–V conversion. The output voltage V_{DAC} is proportional to the output current I_{DAC}

In our case, linear resistors are not available and negative feedback configurations are problematic. Moreover, for display applications the DAC is typically loaded by a diode, hence a diode connected load was used to convert the current in a voltage (TFT M4 in Fig. 7.15).

Of course this kind of load is far from linear and, thus, a differential approach (Fig. 7.15) was used to cancel the even non linearities. The two input devices switch the unity current (corresponding to a single LSB) completely to the one branch or to the other, with one of TFTs M2 always strongly on and the other strongly off. In this way, however, the bias of the drain tail transistor M1 depends on the output voltage, and hence on the input signal, creating non linearities and distortion. Moreover, the switching activity of the input device causes a considerable charge injection to the output nodes due to the large overlap capacitances associated with TFTs (in particular the gate-drain capacitance C_{GD}).

In order to solve these two problems, cascode transistors M3 have been included in the design of the unity cell (Fig. 7.15). Indeed, they increase the output resistance of the cell (without diode load), making the unity current less sensitive to different output voltages, and they reduce the effect of the charge injection from the differential input to the differential output.

These cascode devices allow also an easy solution to apply post-fabrication calibration (which was not exploited here though). Indeed, controlling independently the bias V_{cas} of each binary weighted cell, the current can be adjusted.

Current-Steering DAC Design

To avoid complex control logic, a CS-DAC with unary current source and binary selection has been chosen (Fig. 7.15). In this case, the maximum DNL can be

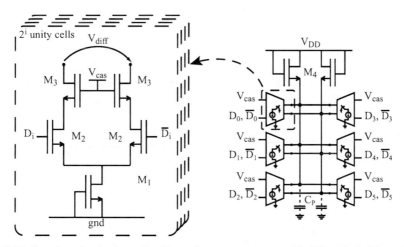

Fig. 7.15 Transistor level schematic of the unity current source and building block schematic of the 6 bit CS-DAC. (© 2012 IEEE. Reprinted, with permission, from D. Raiteri et al. [2])

related to the statistical current error ΔI in the unity current sources of the DAC (Eq. 3.41 in [18]):

$$DLN_{max} = \pm 2^{\frac{n-1}{2}} \frac{\Delta I}{I} \tag{7.10}$$

where n is the number of bits and I the unity current. At a first approximation, the maximum normalized current error $\Delta I/I$ is due to threshold and mobility variations, thus $\Delta I/I > \Delta\mu/\mu$, $\Delta\mu/\mu$ being the maximum normalized mobility variation. In the GIZO technology used for this design, on an area comparable to the one occupied by our DAC, $\Delta\mu/\mu$ has been measured to be 4 % [13]. Based on this information and Eq. (7.10), the DNL_{max} of a 7 bit DAC would be worse than 0.64 LSB. For the sake of compactness and simplicity, a 6 bit resolution was preferred. In this way, half the amount of unity cells and interconnection is required, with a great reduction in area and better hard faults probability. Moreover, from this implementation a DNL better than half LSB would be expected.

In this unipolar technology, like in the one described in Chap. 4, transistors are normally-on. For this reason the implementation of current mirrors is inaccurate (as discussed in Chap. 6), and the unity current cannot be provided by an external reference. Therefore, the current of the unity cell is defined by means of the W/L ratio and threshold of the Zero-Vgs connected device M1. With this approach, the only way to change the unity current after fabrication is by means of the cascode bias V_{cas}. As it will be shown later, the cascode bias is used here to correct the static characteristic of the converter, but it can also be used to define the full scale output current. This value should be chosen considering the application addressed, and the unity current should be scaled accordingly. A larger current could drive more easily the parasitic capacitive load enabling better dynamic performance of the converter. On the other hand, a smaller current would dissipate less power.

Current-Steering DAC Realization

Also for this technology, information on matching and parameter variations was very limited. However, it is reasonable to expect that, due to the large-area nature of the process, in addition to short-distance Gaussian variations, also long-distance gradients affect the process parameters, and thus the matching between the devices (and eventually the DAC linearity).

For this reason, a common centroid criterion was followed to place the 63 unity current sources of the converter. In Fig. 7.16a, the common centroid floor plan is shown. For instance, the third bit controls four unity cells (boxes numbered with 4), which are located symmetrically with respect to centre of gravity of the layout.

In this way, however, unity cells connected to the same input bit can be located quite far and interconnections among them become very long and resistive. Multiple interconnections have been drawn (Fig. 7.16b), to reduce their resistance and to avoid circuit failures due to scratches on the surface of the plastic foil. Indeed, the surface of the circuit can be damaged easily, causing a hard fault if an interconnection is broken (e.g. the white spots in Fig. 7.16b). The final circuit occupies an area of 6.5 × 9.5 mm².

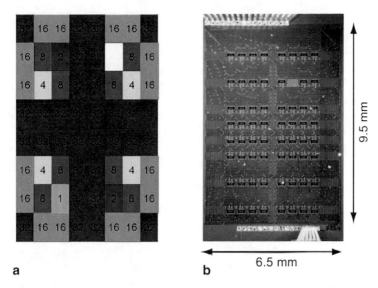

Fig. 7.16 **a** Layout and **b** photograph of the measured current-steering DAC manufactured in amorphous GIZO technology. (© 2012 IEEE. Reprinted, with permission, from D. Raiteri et al. [2])

Measurements

For the sake of simplicity, the DAC has been characterized measuring the differential output voltage V_{diff} with an oscilloscope. The circuit is supplied at $V_{DD} = 3$ V and the measured full scale current is 380 μA.

From the comparison between the single ended and the differential output, shown in Fig. 7.17, it is clear that the differential readout cancels the even non-linearity of the diodes avoiding a strong increase in INL. Without any additional correction the maximum integral non linearity is $INL_{max} = 0.7$ LSB and the maximum differential non linearity is $DNL_{max} = 1.2$ LSB. In the specific measured instance, the largest contribution to the DNL was not caused by the commuting of the MSB, as it would be expected, but by the MSB-1. This unusual result is most probably due to a soft/hard fault on a unity current source connected to the MSB-1.

For this reason, to investigate the effects of possible calibration schemes, the (MSB-1) current has been tuned by reducing the corresponding driving logic '1' voltage from V_{DD} to 2.6 V, as proposed in [19]. The correction was applied in this case directly to the input TFTs M2 of the unity cell. However, in an actual self-calibrating system, a similar correction could be applied to the cascode voltage V_{cas} used to bias the gate of M3. Indeed when the differential switch directs the unity current, one transistor M2 acts like an open circuit and the other shortens the source of the cascode device M3 to the tail transistor M1. In this way the drain-source voltage on the tail device depends on V_{cas} (i.e. $V_{DS,1} = V_{cas} - V_{GS,3}$) and a reduction of V_{cas} allows the reduction of the unity current in the cell. After the adjustment of the current of the (MSB-1) bit, an INL_{max} of 0.2 LSB and DNL_{max} of 0.3 LSB have been measured (Fig. 7.17b).

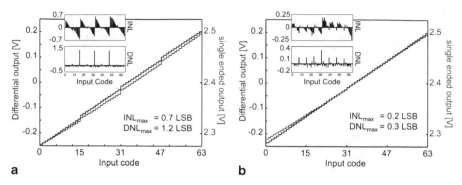

a   b

Fig. 7.17 Measured characteristics of the CS-DAC **a** before and **b** after the MSB-1 adjustment and corresponding INL and DNL in the inset

The output spectra of the differential voltage V_{diff}, measured using a sampling rate $f_{SAM} = 1$ MS/s, are shown in Fig. 7.18a for a 30 kHz and in Fig. 7.18b for a 300 kHz full-scale sinusoidal input signal. The amplitude of the 30 kHz output is about 0.25 V.

A first order roll-off after ~ 80 kHz, due to the capacitive loading offered by the measurement setup ($C_p \sim 100$ pF which is much bigger than the internal parasitics), is observed at higher signal frequencies. Figure 7.19 displays the measured

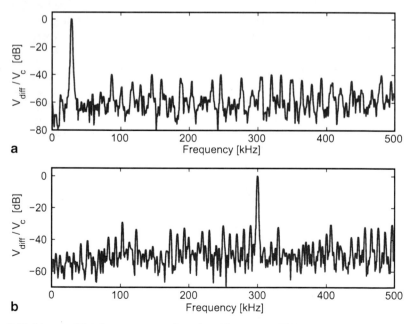

a

b

Fig. 7.18 Measured DAC output spectra for a signal frequency of 30 kHz and 300 kHz normalized to the output signal amplitude (V_c). Sampling rate $f_{SAM} = 1$ MS/s. (© 2012 IEEE. Reprinted, with permission, from D. Raiteri et al. [2])

Fig. 7.19 Measured Spurious Free Dynamic Range (SFDR) as a function of the input frequency for different sampling frequencies f_{SAM}. (© 2012 IEEE. Reprinted, with permission, from D. Raiteri et al. [2])

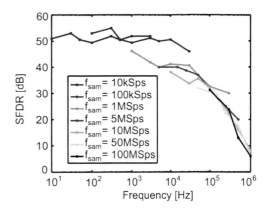

SFDR (Spurious Free Dynamic Range) of the DAC output, shown for sampling rates f_{SAM} going from 10 kS/s to 10 MS/s. An SFDR better than 50 dB can be achieved for input frequencies up to 10 kHz and better than 30 dB for input frequencies up to 300 kHz.

Another way to estimate the DAC dynamic behavior, especially relevant to the application of the column signal generation in displays, is the time-domain response. The transient of V_{diff}, when a single digit of the input word switches to '1' (Fig. 7.20), settles in about 4 µs. Considering for instance QVGA applications (display matrix of 320×240 pixels at 30 frames per second), the CS-DAC is then suitable for a 20x multiplexing factor among pixels of the same row.

Considering the DAC linearity and speed, this component could also be used to design a 6 bit resolution SAR converter with a conversion rate in the order of 100 kS/s (allowing 6 µs for comparison and logic). A benchmarking among the GIZO DAC and organic [19], a-Si [14] and LTPS [15] state of the art DACs for large-area applications is presented in Table 7.1.

Fig. 7.20 Measured time response to a single bit step. For the lower line, only the LSB was switched from '0' to '1', while for the largest step only the MSB changes. (© 2012 IEEE. Reprinted, with permission, from D. Raiteri et al. [2])

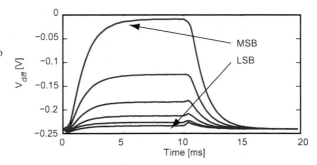

Table 7.1 Benchmark among the GIZO current-steering DAC and state-of-the-art DACs for display applications

	[19]	[14]	[15]	This work
Technology	p-type OTFT	a-Si:H	LTPS	GIZO
DAC implementation	Current steering	Capacitive	Resistive	Current steering
DAC resolution	6b	7b	8b	6b
Area	2.6×4.6 mm^2	8 mm^2	0.12×3 mm^{2a}	6.5×9.5 mm^2
Voltage supply	3.3 V	20 V	N.A.	3 V
Full scale output current	79 µA	N.A.	N.A.	380 µA
DNL$_{max}$	0.69 LSB	0.6 LSB	1 LSB	0.3 LSB
INL$_{max}$	1.16 LSB	1 LSB	>«1 LSB	0.2 LSB
SFDR$_{max}$	32 dB	N.A.	N.A.	55 dB
Max. sampling rate	100 kS/s	N.A.	N.A.	10 MS/s
Settling time	N.A.	~40 µs	~3 µs	~4 µs

[a] Estimated from Fig. 7 of [15]

7.3 VCO-Based Analog to Digital Converter

The two main causes of non-linearity in several ADC topologies are a variable offset among different comparators, and a nonlinear reference. In case of flash ADCs both of these contributions need to be taken into account, while for SAR ADCs only the linearity of the reference source plays a role on the overall static characteristic. As shown in Chap. 5, a solution that overcomes also the issue of a linear voltage reference can be found in the integrating ADCs. In this case, the design effort is mainly focused on the linearity of the integration and on the realization of a low-complexity logic. This approach brought us to obtain the smallest large-area converter reported in literature, and to achieve the highest linearity to date without any analog calibration technique, just applying an offset-gain correction to the output data.

7.3.1 The Proposed VCO-Based ADC

The integrating AD converter proposed is a VCO-based ADC where the issues caused by the mismatch are avoided basing the resolution on the conversion time, and the conversion linearity on the linearity of the voltage-frequency characteristic of a single Voltage Controlled Oscillator (VCO). Indeed, the conversion principle is based on amplitude-to-frequency conversion first and on a frequency-to-phase conversion afterwards. Moreover, in ultra-low cost applications, like smart sensors integrated in packaging of food or pharmaceuticals (see Chap. 2), the system complexity needs to be low in order to achieve high circuit yield against hard and soft faults. For this reason, digital post-processing is left to the base station that retrieves the digitalized data, while the proposed ADC is mostly digital, and exploits only one VCO and a 10 bit ripple counter (Fig. 7.21a).

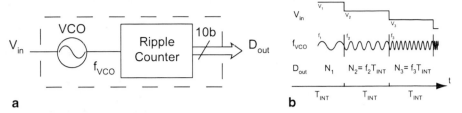

Fig. 7.21 **a** Building block schematic of the implemented VCO-based ADC and **b** conversion principle. (© 2013 IEEE. Reprinted, with permission, from D. Raiteri et al. [3])

Converter Analysis

The VCO-based ADC is made of only two building blocks (Fig. 7.21a) reducing drastically the circuit complexity and the hard/soft fault probability.

The conversion principle is explained visually in Fig. 7.21b. The input signal controls the oscillation frequency of the voltage controlled oscillator f_{VCO}. The ripple counter is used as a digital integrator to count the number of full periods during the integration time T_{INT} and the count number gives the digital output word. Since the linearity is not based on matching, in first approximation it is not affected by process variations, which can only cause offset and gain errors. The size of these errors can be estimated by the base station that receives the raw data and adjusts, for instance, the integration time to correct the gain error. Another solution is to preserve a constant integration time (for instance generated by a local oscillator) and to convert, in the beginning, an input signal equal to the extremes of the input range. In this way the base station can easily extract for the specific VCO-based ADC the right offset and gain errors, which are trivial to correct in the digital domain by the base station. It is important to notice that the local oscillator generating the integration time does not need to be long-time stable, if the base station applies the gain and offset corrections frequently enough.

Converter Design and Simulation

This section deals with the design of the VCO-based ADC. First, the voltage controlled oscillator and the linear transconductor used to control the oscillation frequency are presented. The logic style used for our digital circuits will be extensively studied in Chap. 8, thus only a short discussion on the ripple counter is given here.

Voltage Controlled Oscillator

The VCO designed for our converter comprises 7 fixed-delay inverting stages and a non-inverting tunable delay cell (Fig. 7.22a). The tunable delay is much larger than the delay of the seven inverters, and hence it dominates the oscillation output

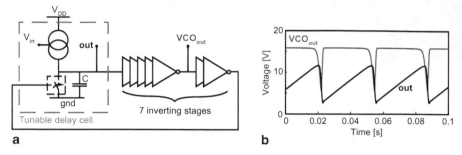

Fig. 7.22 **a** Building block schematic of the voltage controlled oscillator and **b** simulated temporal behavior of the output of the tunable delay cell (*black*) and the output of the VCO (*gray*). (© 2013 IEEE. Reprinted, with permission, from D. Raiteri et al. [3])

frequency. The fixed-delay chain made by the seven inverters accomplishes two tasks. First, it defines the temporal width of the negative pulse (that should last enough to reset the output node 'out'); second, it squares the saw tooth signal provided by the tunable delay cell in order to trigger the clock input of the ripple counter with a squared signal.

After the capacitor C has been reset, a current proportional to the input voltage V_{in} is provided by a linear transconductor, generating a linear ramp at the output of the delay cell (Fig. 7.22b). When the linear ramp reaches the threshold voltage of the first inverter of the chain of seven inverters, an edge starts propagating through the fixed delay line. After the fifth inverter the negative pulse triggering the counter is generated and after the seventh the capacitance C is reset by the switch. At this point a second edge starts propagating through the delay line, the trigger is brought back to the logic state "high", and the switch opens allowing the capacitor to charge up again.

The pull-down of the non-inverting element is performed connecting the output node to ground by means of a low impedance, hence it is very fast and introduces a negligible delay compared to the one of the delay cell. On the other hand, the pull-up time depends on the load capacitance (C = 500 fF) and on the small output current provided by the transconductor. Therefore, the negative pulse width is independent of the input signal, and the maximum and minimum output frequencies of the oscillator mainly depend on the maximum and minimum currents provided by the transconductor.

Linear Transconductor

The transconductor used in this circuit improves the one presented in the previous chapter: it requires only two OTFTs instead of five and achieves better performance in terms of linearity and output resistance. Another feature of this transconductor is that the input signal is fed through the top-gate, which provides extremely low capacitance (the top gate dielectric is 1.4 μm thick), minimizing the sensor capacitive load to just about 34 fF.

Fig. 7.23 Schematic of
the transconductor used in
the VCO. (© 2013 IEEE.
Reprinted, with permission,
from D. Raiteri et al. [3])

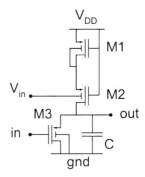

In this transconductor (Fig. 7.23), the current source M1 is used to bias the common gate TFT M2. Varying the top-gate voltage of the input transistor M2, its threshold voltage changes linearly, as shown in Chap. 4, and so does the gate-source voltage of M2 (biased with an almost-constant current). This variation affects only the source voltage of M2 since its gate is connected to the supply, hence a linear relation is created between the source-drain voltage of M1 and V_{in}. The channel length modulation of M1 will then generate a linear variation of the current in M1 (and thus of the output current) with V_{in}. The connection of the top-gate of M1 to its drain adjusts the value of the output resistance, further improving the linearity. One should also remind that our p-type OTFTs are normally-on, and thus the source voltage of M2 can indeed be lower than its gate bias, V_{DD}, while keeping correct circuit functionality.

Ripple Counter

The digital counter performs the integration that converts the analog input frequency into a digital output; for this reason, the number of stages of this circuit determines the maximum resolution of the circuit. In our case, a 10-stage ripple counter was designed employing 10 data flip flops (DFF), hence 10 bit is also the maximum resolution achievable when the correct integration time is allowed for the integration. It is worth noting, however, that the effective number of bits (ENOB) can be less, due to thermal noise in the voltage controlled oscillator that causes jitter at the input of the counter, and to the non-linearity of the VCO.

The building block schematic of the ripple counter is shown in Fig. 7.24. The output signal of the VCO triggers the first stage, while all the others are triggered by the output of the previous one. From a system point of view, the only requirement on the ripple counter is the speed, since the counter has to be able to increment its output of one unit at the maximum frequency achievable by the VCO. Due to the slow VCO output the ripple counter speed is more than sufficient and this counter structure, which enables the smallest amount of logic for a given maximum count, could be used.

The logic used to design this circuit is analyzed in depth in the Chap. 8, as well as the schematic of the DFF.

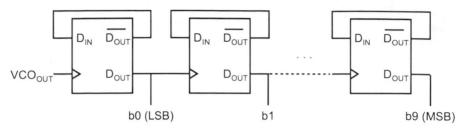

b0 (LSB)   b1   b9 (MSB)

Fig. 7.24 Building block schematic of the ripple counter in use

Converter Realization

The photograph of the complete converter is shown in Fig. 7.25. Thanks to the simplicity of the circuit topology, the analog circuitry is reduced to the minimum and most of the area is occupied by the digital counter.

The whole converter, including contact pads, occupies less than $3.5 \times 5.7\,\mathrm{mm}^2$. This is by far the smallest converter realized with large-area technologies (see Table 7.2). In fact, the area is $19.4\,\mathrm{mm}^2$, almost 14 times smaller than previous works [20] that reaches lower effective resolution and uses the same TFT technology.

Fig. 7.25 Photograph of the VCO-based ADC on foil. (© 2013 IEEE. Reprinted, with permission, from D. Raiteri et al. [3])

Table 7.2 Benchmark among the VCO-based ADC and state-of-the-art ADCs manufactured with low-temperature technologies on foil

	Architecture	Resolution [bit]	SNR [dB]	DNL [LSB]	INL [LSB]	Current consumption [µA]	Area [mm²]
[20]	ΣΔ M	4.1	26.5	N.A.	N.A.	100	260
[9]	SAR (C-2C DAC)	6	N.A.	2.6[a]	3[b]	1.2	616
[3] This work	VCO-based	6	48	0.6	1	2.4	19.4

[a] ENOB evaluated from the SNR
[b] Of the C-2C DAC before calibration

Converter Measurements

In this section, the experimental characterization of the linear transconductor is shown first, and the measurement results on the VCO are shown afterwards. The VCO characteristics are then related to the static performance of the full converter. Unfortunately an error in the layout of the ripple counter hampered the measurement of the whole system and an external counter has been used instead. However, the functionality of the same data flip-flop used in the counter is experimentally demonstrated in Chap. 8.

Linear Transconductor

With respect to the topology shown in the Chap. 6, higher linearity and higher output resistance have been reached by trading them off with the tunability to cope with parameter variability. To this extent, Fig. 7.26 shows the measurements of 54 transconductors manufactured on the same foil and affected by the typical variability that cannot be, in this case, corrected. The maximum current, for instance, varies of about one order of magnitude among all samples, causing gain and offset variations in the ADC characteristic. These variations however can be corrected in the digital domain on the digital output. In fact, offset correction could be easily performed using an up/down counter, and the gain can be adjusted by means of a digital divider. This circuit anyway demands too complex digital logic to be integrated on chip with our technology, as this would introduce yield and area issues. An effective solution to variability would be thus to convert first the two rail voltages, and then the voltage of interest, letting the base station correct the offset and gain after transmission of these three digital values.

Following this approach, the only residual linearity error comes from the integral non-linearity (INL). The variability of the linearity is remarkably smaller than the absolute variation of the current. This can be seen in the graph showing the normalized current linearity error in Fig. 7.27. In this plot, the error current of each transconductor is first evaluated as the difference between the measured current and the linear interpolation through the extremes. Then it is divided by the current corresponding to an input voltage equal to half the input range. The outcome of this elaboration shows a systematic behavior of the error, due to the nonlinear relation between the output current and the drain-source voltage applied. This non linearity is due to the channel length shortening which is less sensitive than other process parameter to variations, as for instance the threshold voltage or the mobility. For this reason, the variability of the normalized current linearity error (Fig. 7.27) among the various transconductors is much less than the variation of the absolute current.

The thick lines in Figs. 7.26 and 7.27 represent the measurements performed on the transconductor closest to the VCO that was fully characterized. The transconductance for a DC input voltage V_{in} of 10 V, and in saturation regime, is ~8.5 pA/V. The output resistance for $V_{in} = 5$ V is ~4.8 TΩ.

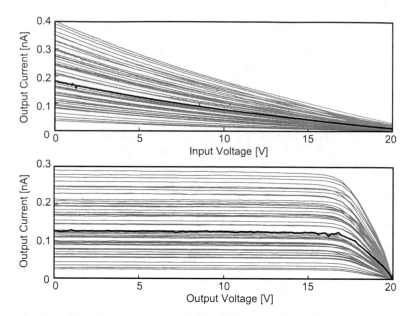

Fig. 7.26 a Transfer and **b** output characteristics of 54 transconductors measured: the *thick lines* are measured from the transconductor closest to the VCO characterized. (© 2013 IEEE. Reprinted, with permission, from D. Raiteri et al. [3])

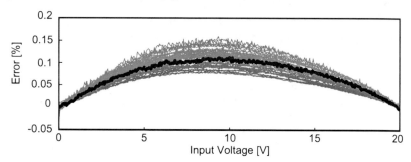

Fig. 7.27 Normalized current linearity error for the 54 samples measured. The thick line is measured from the transconductor closest to the VCO characterized

Voltage Controlled Oscillator

Figure 7.28 shows an example of the output signal coming from the VCO and triggering the counter.

For the measured sample, the VCO output frequency goes from 4.15 Hz (red line) to 37.65 Hz (blue line) for a rail-to-rail V_{in} variation (0–20 V).

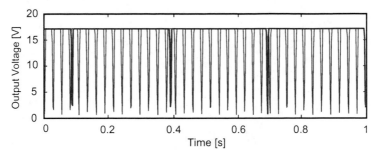

Fig. 7.28 Measured output of the VCO for maximum (*blue line*) and minimum (*red line*) input voltage, for respectively $V_{in}=20$ V and $V_{in}=0$ V. (© 2013 IEEE. Reprinted, with permission, from D. Raiteri et al. [3])

The Full Converter

As shown at the beginning of this section, the digital output can be obtained integrating the VCO output for a fixed time T_{INT}. Since integration is a linear transformation, the linearity of the final converter can be estimated by the linearity of the VCO voltage-frequency characteristic (Fig. 7.29).

A longer integration time improves the resolution of the conversion, as it reduces the effects of thermal noise. In our case, an SNR between the VCO period and its jitter of ~48 dB was measured, corresponding to an ENOB due to noise only: $ENOB_{noise}=7.7$ bit. As the resolution N of the converter depends on the frequency range f_{span} of the VCO and on the conversion time T_{INT}, the minimum conversion time needed to exploit the noise-limited SNR is $T_{INT}=6.2$ s. The INL and DNL of the converter are plotted in Fig. 7.30 with the resolution of the converter fixed to 6 bits and for a rail-to-rail input range. The worse-case INL is 1 LSB, while the maximum DNL is 0.6 LSB. Measurements were performed using an external counter, due to the layout problems affecting the integrated one.

In the realized chip, ten DFFs, designed according to the logic style that will be discussed in Chap. 8, are connected to implement the 10 bit ripple counter that was integrated with the VCO. To guarantee the correct behavior of the circuit, the counter needs to be functional up to the maximum frequency that can be generated by the VCO, which is, in this case, ~40 Hz.

The VCO draws a current of ~300 nA, while the current consumption of the counter (based on the measurements performed on the shift register shown in Sect. 8.1.4) is almost ~2.1 μA. The SNR and linearity demonstrated improve the state of the art [9, 20], (Table 7.2) and could be obtained exploiting the linearity of the proposed transconductor, rather than the poor matching of organic technologies, together with digital-only offset and gain corrections.

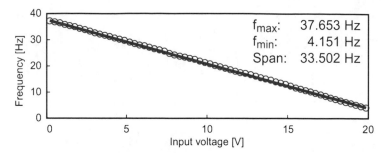

Fig. 7.29 Measured voltage-frequency characteristic of the VCO characterized. *Circles* show the measured data while the *red line* is the linear fit through the extremes. (© 2013 IEEE. Reprinted, with permission, from D. Raiteri et al. [3])

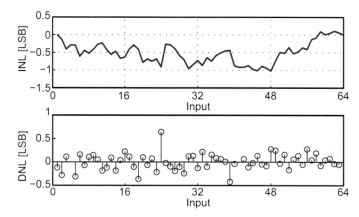

Fig. 7.30 Measured INL and DNL of the converter at 6 bit resolution exploiting a rail-to-rail input range. (© 2013 IEEE. Reprinted, with permission, from D. Raiteri et al. [3])

7.4   Conclusions

In this chapter, the design of data converters in large-area unipolar technologies was discussed. In particular, a latched comparator was first analyzed, since it is needed in all ADC architectures of the type shown in Fig. 5.2 and forms the fundamental building block in a flash ADCs. Indeed, in flash converters, a large number of comparators (exponentially increasing with the resolution N, i.e. 2^{N-1}) compare the input signal to the same number of reference voltages. This topology targets fast applications, hence a fast comparison is required. For this purpose, the design started from a standard latch design and its time response was improved approximately by ten times (from 50 to 5 μs), achieving the limits imposed by the technology. Differently from other comparators manufactured with large-area technologies [21], the proposed one also provides rail-to-rail outputs with a great benefit for the circuitry following. Some design choices can lower the load for the circuit driving the latch. To this end, exploiting the double-gate feature of the technology, the input capacitance was reduced more than two orders compared to the basic latch implementation. This result makes the latch compatible with the differential parametric amplifier shown in the previous chapter: a perfect solution to reduce the input referred offset of the latch and achieve higher ADC linearity.

To improve the converter linearity, different ADC topologies can be employed. For instance, SAR converters adopt only one comparator. Hence, the non linearities only depend on the reference generated by the DAC. For this reason, a current-steering DAC was designed, that can be employed both for AD conversion in SAR ADCs or as a pixel driver in display applications. The final design demonstrates good linearity (of about 6 bit) and high speed, due to the high mobility of the metal-oxide technology exploited. The better linearity, compared to other current-steering DACs demonstrated in large area electronics, is achieved thanks to a differential approach, to a common centroid layout, and to a technology providing better uniformity and thus matching.

Finally, an integrating ADC was discussed. Indeed smart-sensors manufactured with large-area electronics mainly target quasi-static applications and this converter topology provides, for such applications, better linearity and less complexity. A VCO-based ADC was exploited that consists of only two building blocks: a VCO and a digital counter. The proposed VCO is based on a linear transconductor that improves linearity and output resistance of the one discussed in Chap. 6. The proposed ADC achieves better linearity than other ADCs reported in literature and occupies a considerably smaller area, with beneficial effects on the yield.

References

1. D. Raiteri, F. Torricelli, P.V. Lieshout, A.H.M.V. Roermund, E. Cantatore, A synchronous rail-to-rail latched comparator based on double-gate organic thin-film-transistors. *IEEE European Solid-State Ciruait Conference*, pp. 141–144, 2012
2. D. Raiteri et al., A 6b 10MS/s current-steering DAC manufactured with amorphous Gallium–Indium–Zinc–Oxide TFTs achieving SFDR > 30dB up to 300 kHz. *IEEE International Solid-State Circuits Conference*, pp. 314–316, 2012
3. D. Raiteri, P.v. Lieshout, A.v. Roermund, E. Cantatore, An organic VCO-based ADC for Quasi-Static signals achieving 1LSB INL at 6b resolution. *IEEE International Solid-State Circuits Conference*, pp. 108–109, 2013
4. B. Razavi, B.A. Wooley, Design techniques for high-speed, high-resolution comparators. IEEE J. Solid-State Circuits **27**(12), 1916–1926 (1992)
5. M.J.M. Pelgrom, H.P. Tuinhout, M. Vertregt, Transistor matching in analog CMOS applications. *International Electron Devices Meeting*, pp. 915–918, 1998
6. D. Raiteri, F. Torricelli, E. Cantatore, A.H.M.V. Roermund, A tunable transconductor for analog amplification and filtering based on double-gate organic TFTs. *IEEE European Solid-State Circuits Conference*, pp. 415–418, 2011
7. D. Raiteri, E. Cantatore, A.H.M.V. Roermund, *A Digital Buffer for AC Measurements of Unipolar Organic TFT Circuits with Picoprobes*. (ICT.OPEN, 2012)
8. W. Xiong, Y. Guo, U. Zschieschang, H. Klauk, B. Murmann, A 3-V, 6-Bit C-2C digital-to-analog converter using complementary organic thin-film transistors on glass. IEEE J. Solid-State Circuits **45**(7), 1380–1388 (2010)

9. W. Xiong, U. Zschieschang, H. Klauk, B. Murmann, A 3 V 6b successive-approximation ADC using complementary organic thin-film transistors on glass. *IEEE Int.ernational Solid-State Circuits Conference*, pp. 47–49, 2010
10. K. Nomura et al., Room-temperature fabrication of transparent flexible thin-film transistors using amorphous oxide semiconductors. Nature **432**(7016), 488–492 (2004)
11. H. Ohara et al., 4.0-in active-matrix organic light-emitting diode display integrated with driver circuits using amorphous In–Ga–Zn-Oxide thin-film transistors with suppressed variation. Jpn. J. Appl. Phys. **49**(10), 03CD02 (2010)
12. H. Yin et al., Program/erase characteristics of amorphous Gallium Indium Zinc Oxide nonvolatile memory. IEEE Trans. Electron Devices **55**(8), 2071–2077 (2008)
13. A.K. Tripathi et al., Low-voltage gallium-indium-zinc-oxide thin film transistors based logic circuits on thin plstic foil: Building blocks for radio frequency identification application. Appl. Phys. Lett. **98**(16), 162102 (2011)
14. A. Dey, H. Song, T. Ahmed, S.M. Venugopal, D.R. Allee, Amorphous silicon 7 bit digital to analog converter on PEN. *IEEE Custom Integrated Circuits Conference*, pp. 1–4, 2010
15. Y. Park, D. Kim, K. Kim, An 8b source driver for 2.0 in full-color active-matrix OLEDS made with LTPS TFTs. *IEEE International Solid-State Circuits Conference*, pp. 8–10, 2007
16. Y.-H. Tai, H.-L. Chiu, L.-S. Chou, C.-H. Chang, Boosted gain of the differential amplifier using the second gate of the dual-gate a-IGZO TFTs. IEEE Electron Device Lett. **33**(12), 1729–1731 (2012)
17. C. Zysset, N.S. Münzenrieder, T. Kinkeldei, K.H. Cherenack, G. Tröster, Indium–gallium–zinc–oxide based mechanically flexible transimpedance amplifier. Electron. Lett. **47**(12), 691–692 (2011)
18. F. Maloberti, *Data Converters*. (Springer, Dordrecht, 2007)
19. T. Zaki et al., A 3.3 V 6-bit 100 kS/s current-steering digital-to-analog converter using organic P-type thin-film transistors on glass. IEEE J. Solid-State Circuits **47**(1), 292–300 (2012)
20. H. Marien, M.S.J. Steyaert, E. van Veenendaal, P. Heremans, A fully integrated ΔΣ ADC in organic thin-film transistor technology on flexible plastic foil. IEEE J. Solid-State Circuits **46**(1), 276–284 (2011)
21. H. Marien, M. Steyaert, N.V. Aerle, P. Heremans, A mixed-signal organic 1 kHz comparator with low VT sensitivity on flexible plastic substrate. *IEEE European Solid-State Circuit Conference*, pp. 120–123, 2009

Chapter 8
Circuit Design for Digital Processing

Digital circuits are extensively used in any integrated system, sometimes to assist analog or mixed-signal functions and sometimes to perform logical operations. In both cases, logic gates should achieve high gain and large noise margin in order to provide robustness and enable high circuit yield. In this chapter, a new logic style is proposed that, compared to known approaches, improves gain, noise margin, and yield. Furthermore, exploiting only control signals within the power supply range, the proposed logic enables the integration of control circuits to automatically correct for process variations and aging. The content of this chapter has been published in [1].

8.1 The Proposed Positive-Feedback Level Shifter Logic

The detailed analysis of the state-of-the-art logic styles shown in Chap. 5 allows to point out a few issues which concern present digital logic manufactured with TFTs on foil. First of all, the gain and the symmetry of the input–output characteristic should be large enough to guarantee the necessary noise margin. Next, process variations should not compromise the functionality of the logic. As this is not the case in large-area low-temperature technologies, a post-processing tuning control is desirable.

To enable practical applications of large-area electronics on foil, the post-processing tuning should be performed automatically by the circuit, and not use arbitrary-large voltage sources in the laboratory. To enable the integration of such control circuitry, a Positive-feedback Level Shifter (PLS) logic is proposed. This logic only exploits control signals that are within the power supply range, to control the symmetry of the characteristic and achieve high noise margin despite the considerable process variations.

© Springer International Publishing Switzerland 2015   111
D. Raiteri et al., *Circuit Design on Plastic Foils,* Analog Circuits and Signal Processing,
DOI 10.1007/978-3-319-11427-9_8

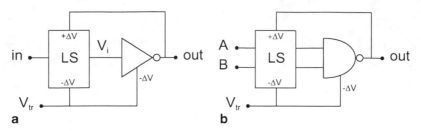

Fig. 8.1 Building block schematic of **a** a PLS inverter and **b** a PSL NAND gate

8.1.1   PLS Logic Analysis

For our PLS logic, a two-stage topology is exploited where the output stage is a double-gate enhanced logic gate with Zero-Vgs connected load (see Chap. 5), and the first stage is a tunable level shifter LS (Fig. 8.1). As explained in Chap. 5, in the dual-gate enhanced output stage the threshold between the input logic "zero" and input logic "one" is controlled by a tuning input. If the voltage applied to this input ($-\Delta V$) increases, the threshold voltage becomes lower. In the level shifter, two inputs control the amount of shift between the input and the output signal: when the control inputs increase, one ($-\Delta V$) increases the shift and lowers the threshold voltage V_{trip}, while the other ($+\Delta V$) decreases the shift and increases V_{trip}. The output stage determines the function of the whole logic gate: in Fig. 8.1a is shown the block-level schematic of our PLS inverter, while in Fig. 8.1b is shown the schematic of a PLS NAND. These two gates were designed and used in our digital circuits: any other logic function can be implemented based on these two, as known from the Boolean algebra.

In these circuits, V_{tr} is a control voltage that varies the position of the logic threshold, and thus the symmetry of the characteristic. It is applied in both the shifter and the inverter so that, for increasing V_{tr}, the threshold between login "zero" and logic "one" at the input is shifted to lower input voltages. In this way, the effectiveness of the trip-point control is strongly improved with respect to the double-gate enhanced logic, and the transfer characteristic of the inverter can be shifted almost over the full input range, while still exploiting a control voltage within the supply rails.

The second input of the level shifter, the one decreasing the shift, is connected to the output of the inverter. In this way, a positive feedback is created around the trip point that drastically boosts the gain of the inverter. Let us consider, for instance, a transition low–high of the input, the normal operation of the level shifter and of the inverter respectively increases the intermediate voltage V_i (the level-shifted version of the input voltage) and brings the output down. As in the PLS logic the inverted output is connected to the $-\Delta V$ input of the level shifter, the voltage difference between V_i and the input voltage V_{in} further increases when the output drops. The voltage V_i raises more than what the input alone would be able to do, making in turn

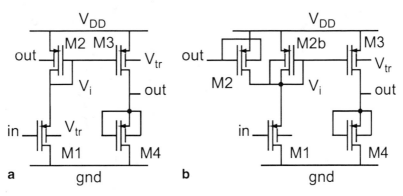

Fig. 8.2 Two different embodiments of the PLS inverter: **a** one that achieves wider trip point tunability, **b** another that achieves larger gain and noise margin. (© 2014 IEEE. Reprinted, with permission, from D. Raiteri et al. [1])

the output drop more. A positive feedback loop is thus generated and a considerable increase of the gain is achieved.

In order to describe the features of the PLS logic, the PLS inverter will be considered; however, all the concepts that follow directly apply also to the NAND gate.

Two transistor-level implementations of the schematic in Fig. 8.1a are shown in Fig. 8.2. The output inverter is embodied by transistors M3 and M4, while the level shifter is formed by M1 and M2. In the embodiment of Fig. 8.2b, the combination of transistors M2 and M2b allows a higher gain for the positive feedback loop.

To analyze the circuit behavior, it is first investigated qualitatively how the control voltage V_{tr} shifts the characteristic, and then how the positive feedback affects the inverter gain. Equation (4.10) shows that the voltage applied to the top-gate of TFTs M2 and M4 causes a variation of their threshold voltage and thus of $V_{SG,1}$ and $V_{SG,3}$. Moreover, the variation of $V_{SG,3}$ also affects the bias current in the level shifter, thus reinforcing the effect on the shift between V_{in} and V_i.

In order to evaluate if the gain in the positive feedback loop is larger than 1, and thus sufficient to guarantee regeneration, we can assume that V_{in} has a fixed value, open the circuit on the gate of the output pull-up transistor M3, place a small-signal voltage source v_s on this node, and evaluate the gain between v_s and v_i.

As shown by the small-signal equivalent circuit in Fig. 8.3, the pull-down TFT M4 and the source follower M1 can be replaced with an equivalent resistor. Indeed the first TFT works in saturation with a constant overdrive ($V_{GS} = V_{TGS} = 0$ V) and can be replaced with its output resistance $R_4 = r_{0,4}$, while the second one has both gate and top-gate connected to bias points, and can therefore be replaced with a resistor equal to:

$$R_1 = \frac{1}{g_{m,G1} + g_{m,TG1}} \qquad (8.1)$$

Fig. 8.3 Simplified small-signal equivalent circuit for the evaluation of the loop gain G_{loop}. (© 2014 IEEE. Reprinted, with permission, from D. Raiteri et al. [1])

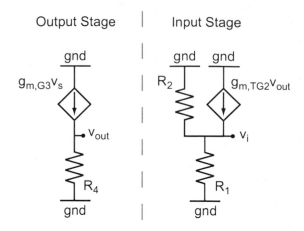

where $g_{m, G1}$ and $g_{m, TG1}$ are the transconductances associated respectively to the gate and to the top-gate. Also the gate and the top-gate transconductance of M2 play a role in the small-signal loop gain. The gate is connected to the drain and thus the gate transconductance can be replaced by a resistor $R_2 = 1/g_{m, G2}$. The effect of the top-gate is modeled as a voltage controlled current source with transconductance $g_{m, TG2}$. Using the simplified equivalent circuit in Fig. 8.3, the loop gain G_{loop} can be evaluated as:

$$G_{loop} = \frac{v_i}{v_s} = -g_{m,G3}r_{0,4}\left(-\frac{g_{m,TG2}}{g_{m,G1} + g_{m,TG1} + g_{m,G2}}\right) \qquad (8.2)$$

Considering that both M1 and M2 work in saturation, since they have the same dimensions ($W_1/L_1 = W_2/L_2 = 10\ \mu m/5\ \mu m$) and are biased by the same current, can be assumed $g_{m, G1} = g_{m, G2}$. Furthermore, $g_{m, TG} = \eta\, g_{m, G}$ due to the different impact of the gate and the top-gate voltages on the channel current (Eqs. (4.9) and (4.10)). Therefore, Eq. (8.2) can be rewritten as:

$$G_{loop} = \frac{\eta}{2 + \eta}g_{m,G3}r_{0,4} \qquad (8.3)$$

In order to take effective advantage of a positive feedback, the loop gain G_{loop} should be larger than 1, which can be easily achieved increasing the channel width of M3 or the channel length of M4.

   If the loop gain is larger than one, the positive feedback also causes hysteresis, as it will be clear from the measurements. Indeed, at the beginning of the low–high transition of the input, the voltage applied to the top-gate of M2 is high: the small current in M2 causes a small shift between V_{in} and V_i, thus the trip point of the inverter is slightly closer to V_{DD}. On the other hand, when the input transition is high–low, M2 is more conductive in the beginning: the shift between V_{in} and V_i is larger and the trip point is closer to ground. However this hysteresis is not detrimental, but beneficial for the noise margin of the single inverter (Sect. 8.1.4). Some issues are related to the combination of a large hysteresis loop together with large process variations, as explained in Sect. 8.1.4.

8.1.2 PLS Inverter Design

The choice of the dimensions of each transistor can be investigated now, being aware of the main mechanisms taking place within the PLS inverter. Our assumption is that, in the ideal case, the trip point should be around $V_{DD}/2$ with a control voltage $V_{tr} = V_{DD}/2$. Also the fan out of the inverter is assumed to be minimum, thus the minimum channel width will be chosen.

Considering the output stage, it is clear that the W/L ratio of M3 has to be much smaller than that of M4. Indeed, when a low output voltage is required, both M3 and M4 have Zero-Vgs (M4 because of the connection; M3 because V_i equals V_{DD}) and M4 needs to drive much more current than M3 to effectively pull down the output node. On the other hand, when the output has to be high, M3 can take advantage of a source-gate voltage as large as approximately V_{DD}, which results in a much larger current than the one provided by M4. For this reason, in gates with minimum fan out (i.e. designed to drive only one other logic gate), M3 is designed as minimum size TFT and M4 is chosen to ensure a good logic low output. For this reason, the final dimensions are $W_4/L_4 = 700$ μm/5 μm for M4, and $W_3/L_3 = 10$ μm/5 μm for M3. A final remark is that the product $g_{m, G3} r_{0,4}$ must be large enough to ensure a positive feedback loop (Eq. 8.3). If this would not be the case, a longer channel for M4 should be chosen.

The level shifter devices can be designed with the same aspect ratio for both M1 and M2. Indeed, the amount of DC shift is independent of the absolute dimensions of the devices and even for equal dimensions for M1 and M2, M2 provides enough current to create the required amount of shift between V_{in} and V_i. Moreover, as the node V_i is lightly loaded by the small transistor M3, the current in the input branch does not need to be increased because of speed considerations. For this reason, M1 and M2 are also designed as minimum size TFTs to minimize the area occupied by the PLS inverter.

Obviously, if the fan out of the gate must be increased, all TFT widths can be scaled up, keeping the W/L ratios constant.

8.1.3 Test Foil Design

In order to design a digital circuit with high yield, it is also extremely important to understand how the spread of the technology parameters affects a large number of logic gates.

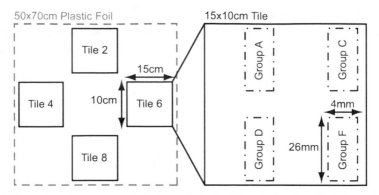

Fig. 8.4 Floor plan of the measured 50×70 cm^2 plastic foil enclosing four 15×10 cm^2 tiles. The four groups included in each tile have 18 repetitions of the PLS inverter. (© 2014 IEEE. Reprinted, with permission, from D. Raiteri et al. [1])

To be able to get such an insight, many inverters were placed on a plastic foil of 50×70 cm^2. The measured inverters can be divided in 16 groups of 18 inverters close to each other. As depicted in Fig. 8.4, the 16 groups are divided on four tiles which are spread on the foil surface and each tile includes 4 groups of inverters. The tiles have a surface of 15×10 cm^2 while each group occupies an area of 4×26 mm^2. For a fair comparison, an equal number of PLS and Zero-Vgs load inverters was designed for measurement on the same foil and in the same regions.

On the same foil, also some analog and digital circuits have been placed: among them a 240-stage shift register made using the PLS logic style (Sect. 8.1.4). This is, to our knowledge, the circuit with highest transistor count ever realized with organic semiconductors (Fig. 8.5).

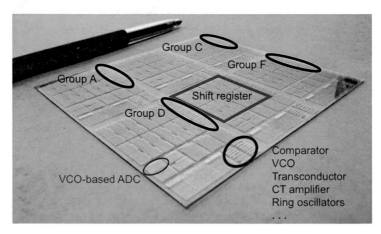

Fig. 8.5 Photograph of one 10×15 cm^2 tile. The PLS test structures and the some of the circuits presented in the previous chapter have been measured on foils like this

Fig. 8.6 Measured transfer
characteristics of a PLS
inverter sweeping V_{tr} from
0 (*rightmost plot*) to 20 V in
steps of 1 V. $V_{tr} = 14$ V returns
for the measured instance the
most symmetric character-
istic and hence the highest
noise margin. The *black*
line shows the characteristic
obtained applying the ideal
bias $V_{tr} = V_{DD}/2 = 10$ V. (©
2014 IEEE. Reprinted, with
permission, from D. Raiteri
et al. [1])

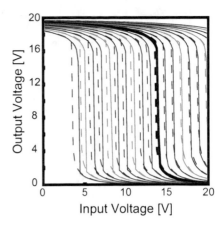

8.1.4   Test Foil Measurements

In this section, are first shown the measured data characterizing a single PLS inverter. Then a statistical analysis will be provided to analyze the improvement that
the PLS approach offers with respect to the most common inverter topology, i.e. the
Zero-Vgs loaded inverter.

Single PLS Inverter

The measured data shown here prove the strong features of the PLS inverter. First,
the trip point tunability and the gain are characterized. Then their relation to the
improved noise margin and other benefits of the PLS topology are discussed.

Trip Point

As shown by the measured inverter transfer characteristics in Fig. 8.6 (in continuous lines for the forward measurements and dashed lines for the backward ones),
using V_{tr} to vary at the same time the threshold of M1 and M3 results indeed in an
effective control of the inverter trip voltage, which can be shifted from about gnd
to V_{DD} while keeping the control voltage V_{tr} within the supply range (the different
transfer characteristics have been measured for V_{tr} sweeping from 0 V, the rightmost
curve, to V_{DD}, the leftmost curve, in steps of 1 V).

It is worth noting that, as a consequence of parameter variations and aging [2],
the designed bias ($V_{tr} = 10$ V) produces, in this particular inverter, a transfer characteristic switching at about 15 V (black line in Fig. 8.6). Applying a control voltage $V_{tr} = 14$ V corrects this asymmetry: the inverter transfer characteristic becomes
perfectly symmetric with respect to the input range.

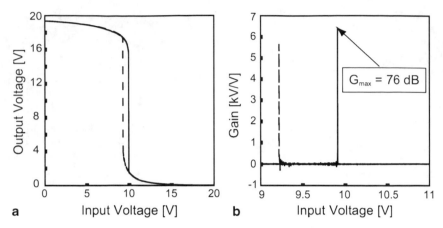

Fig. 8.7 **a** Measured transfer characteristic of the PLS inverter and **b** estimation of the gain. The forward characteristic is represented with a *continuous line*, the backward one with a *dashed line*. (© 2014 IEEE. Reprinted, with permission, from D. Raiteri et al. [1])

Gain

A large gain is mandatory in an inverter to regenerate weak signals and to achieve high noise margin. Organic TFTs suffer from low output resistance and small-signal transconductance, hence the maximum gain of the basic inverter is typically limited to about 20 dB. Moreover, in order to achieve saturation for both the pull-up and pull-down devices at the same time, a large supply voltage is required.

Thanks to the positive feedback, the gain of the PLS inverter increases dramatically. Fig. 8.7a shows the measured forward and backward transfer characteristics of the PLS inverter, while Fig. 8.7b focuses on the transition region, showing a maximum gain of about 76 dB. This value is limited by the 200 μV resolution of the voltage source used at the input; still, to the best of our knowledge, it is more than 26 dB larger than any other organic inverter on plastic foil reported in literature, even if the best previous result exploits ultra-thin dielectric layers (and thus has potentially a larger transconductance and a better output resistance, due to the better electrostatic control that the gate has on the channel) [3].

Noise Margin

The transfer characteristic of the PLS inverter has a very symmetric shape and high gain (Fig. 8.7). Therefore, a good noise margin is expected. Considering, in line with Sect. 5.3, the Maximum Equal Criterion (MEC) in the case of a static characteristic with hysteresis, the noise margin will be given by the side of the largest square that can be inscribed between the forward characteristic and the mirrored backward one. Indeed when the output is high, an unexpected noise source should

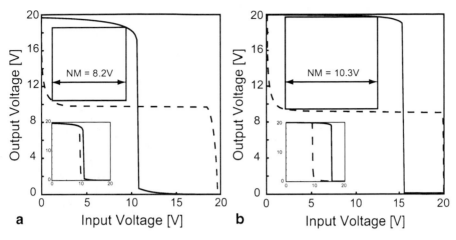

Fig. 8.8 Transfer characteristic and MEC noise margin evaluation for the PLS inverter. **a** in Fig. 8.2a and **b** in Fig. 8.2b. The noise margin is respectively 8.2 V and 10.3 V. The insets illustrate the width of the hysteresis loop. (© 2014 IEEE. Reprinted, with permission, from D. Raiteri et al. [1])

overcome the backward threshold to cause an error; while when the output is low, the noise should overcome the forward threshold.

For this reason, a clockwise hysteresis is not detrimental for the noise margin, but actually beneficial [4–6]. In Fig. 8.8, the noise margins for the two PLS schematics of Fig. 8.2 are shown. For the first PLS topology (Fig. 8.2a) the noise margin is 8.2 V, while the embodiment with higher gain and larger hysteresis (Fig. 8.2b) reaches a noise margin equal to 10.3 V (which is even larger than half the supply). However, if the process variations affecting many different gates are considered, increasing the width of the hysteresis loop reduces the allowed random shift of the characteristic to less than half the difference between the supply V_{DD} and the width of the loop. Hence the probability of soft faults, over a large number of digital gates, increases.

The proposed PLS inverter enables outstanding static performance in case of nominal process parameters and supply conditions. Nevertheless, organic technologies suffer from strong process parameter variations and ageing, and the supply voltage, which can be generated in RFID tags by a rectifier connected to an antenna, can also vary. One of the most important features of the PLS inverter is its tunability, which can effectively counteract all these potential problems.

This flexibility of the PLS design also allows the very same inverter to be functional for a large range of supply voltages. Fig. 8.9 shows the transfer characteristics of a PLS inverter measured for three different supply voltages, $V_{DD} = 5, 10, 20$ V. In all three cases the tuning voltage V_{tr} was kept within the supply range and was set respectively to $V_{tr} = 0, 10$ and 20 V. Even at 5 V supply (which is extremely small for unipolar logic on foil using conventional dielectric layers) the measured PLS inverter still has a noise margin of about 100 mV.

Fig. 8.9 PLS inverter supplied at V_{DD} = 5, 10, 20 V. The control voltage V_{tr} was biased respectively at V_{tr} = 0, 10, 20 V. (© 2014 IEEE. Reprinted, with permission, from D. Raiteri et al. [1])

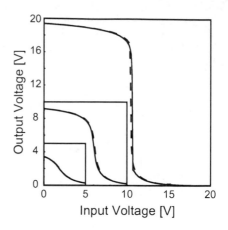

Statistical Inverter Characterization

After the characterization of the single device, we can move further and analyze the data measured on 288 PLS and 288 Zero-Vgs inverters.

The measured data show that all the 288 PLS inverters are functional and can be tuned, choosing V_{tr} individually, to achieve a minimum noise margin of 5.8 V. Two of the 288 Zero-Vgs inverters instead switch for an input voltage around 25 V (for V_{DD} = 20 V), but in this case no voltage control is available to deal with the excessive TFT variability and the soft faults cannot be corrected. The histograms of Fig. 8.10 show the distributions of trip point and noise margin measured on the PLS inverters for two different values of the tuning voltage, namely V_{tr} = 10

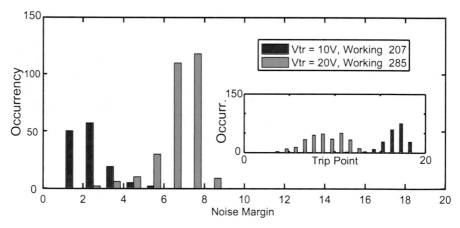

Fig. 8.10 Histograms of trip point and noise margin distributions of 288 PLS inverters measured on the same plastic foil applying the same tuning voltage to all inverters. Two different tuning voltages are shown: V_{tr} = 10, 20 V. (© 2014 IEEE. Reprinted, with permission, from D. Raiteri et al. [1])

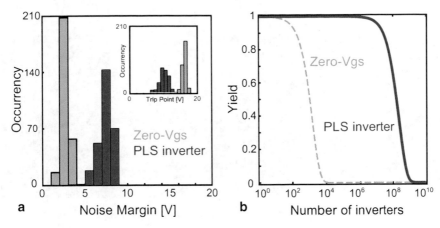

Fig. 8.11 a Histograms of measured trip point (inset) and noise margin distribution of 288 PLS and Zero-Vgs inverters measured on the same plastic foil. For the PLS inverters, the tuning voltage has been chosen suitably for each group of 18 inverters. **b** Yield evaluation based on the measured average and standard deviation of the noise margin (Fig. 8.11a) for PLS (*continuous line*) and for Zero-Vgs (*dashed line*) inverters. (© 2014 IEEE. Reprinted, with permission, from D. Raiteri et al. [1])

and 20 V. Applying the nominal bias $V_{tr} = 10$ V, only 207 inverters show a positive noise margin. Indeed, due to ageing and process parameters variation, the average switching point measured for $V_{tr} = 10$ V is around 17 V instead of half the supply. A large improvement is achieved applying a higher tuning voltage: indeed, forcing $V_{tr} = V_{DD}$, 285 working inverters have been measured. The switching point of the remaining three inverters was shifted for $V_{tr} = V_{DD}$ to negative input voltages, but these inverters work correctly with a lower V_{tr}.

This result shows that a single control voltage is not enough, in our technology, to effectively cope with parameter variations on the entire foil. However a smaller area is likely to be affected by smaller absolute variations, hence the control voltage was also applied separately to each of the 16 groups of inverters (Fig. 8.4). The histograms of noise margin and trip point (inset) evaluated in this way are compared with the Zero-Vgs ones in Fig. 8.11a. Due to the smaller parameter variations within the same group, the majority of the trip points for the PLS inverters is very close to average value $V_{DD}/2$ with a great benefit for the yield.

The average noise margin and its standard deviation can also be used to evaluate the yield of a digital circuit as a function of the number of digital gates in use. As a first-order approximation, it is neglected the effect of the hard faults caused by defects in the TFTs and the interconnections, and it is assumed that the yield is the probability that all gates in digital circuits have a positive noise margin. Following the approach used in [7, 8] a Gaussian probability distribution can be assumed for the noise margin of all the inverters and thus the yield of a digital circuit of an arbitrary number N of inverters can be estimated based on the average and standard deviation of the measured noise margin. Following this approach, the forecasted

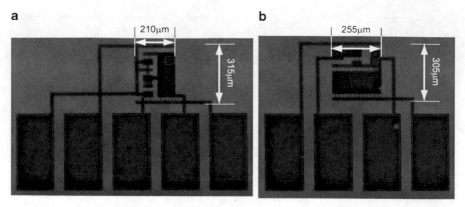

Fig. 8.12 Photographs of the measured **a** PLS inverter and of the **b** Zero-Vgs inverter

yield is plotted in Fig. 8.11b as a function of N. The average and standard deviations of the noise margin are obtained from Fig. 8.11a, both for Zero-Vgs and for PLS inverters. For the latter inverter thus, the case in which the tuning voltage is separately optimized per group of inverters is considered.

To give an example, aiming at a 90% yield from a digital circuit exploiting the Zero-Vgs logic can be found, from the intersection with the dashed line (average noise margin $\mu_{0Vgs} = 2.58$ V, standard deviation $\sigma_{0Vgs} = 0.8$ V), that the number of inverters in the circuit must be smaller than 200. On the other hand, considering the PLS logic (continuous line—average noise margin $\mu_{PLS} = 6.82$ V, standard deviation $\sigma_{PLS} = 1.18$ V), a maximum of N~24 million inverters can be employed. Of course these figures overestimate the actual yield as they do not take into account the yield loss due to hard faults, since, due to the rather small sample size, no hard faults were detected.

However, assuming that the probability of hard faults is proportional to the area occupied by the digital circuit, the PLS inverter would provide an improvement on the Zero-Vgs one also from this perspective, as it occupies only 66,150 μm^2, against the 77,775 μm^2 of the Zero-Vgs one (Fig. 8.12). The Zero-Vgs inverter indeed adopts a much wider output device, since this is the only way to make the static transfer characteristic more symmetric. In the PLS inverter, the symmetry is provided by the level shifter and the output load can be dimensioned only considering the output logic levels. Moreover, the dynamic behavior of the PLS inverter is not negatively affected; indeed the rise time is negligible with respect to the fall time, and the fall time is independent of the load width. In fact, most of the parasitic capacitance is determined by the load transistor (which is Zero-Vgs connected and very wide); hence, the pull-down current and the capacitive load are both proportional to the width of the pull-down transistor. The comparison between Zero-Vgs and PLS logic proves that our circuit approach dramatically improves the yield of digital circuits, decreasing the number of soft faults, and enables functional circuits on foil with an unprecedented transistor count. These outstanding achievements

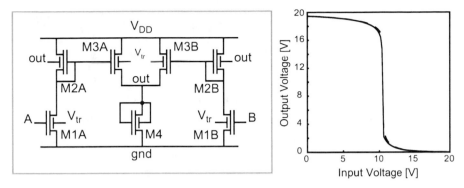

Fig. 8.13 Schematic of a NAND gate in PLS logic. The n-type-only counterpart would employ the devices M3A and M3B in series. (© 2014 IEEE. Reprinted, with permission, from D. Raiteri et al. [1])

Fig. 8.14 Photograph of a 9-stage ring oscillator. (© 2014 IEEE. Reprinted, with permission, from D. Raiteri et al. [1])

can be obtained with a per-group trip point control approach that represents a low-cost solution providing significant yield improvement at the cost of little additional complexity.

Digital Building Blocks and Circuits

The PLS logic style was applied to the basic digital building blocks required to design more complex functions. Ring oscillators have been measured to investigate the speed of the logic and data flip flops (DFF) have been designed to demonstrate a functional 240-stage shift register. However, a NAND gate is required to implement all these digital functions. For this reason, a 2-input NAND gate was also designed following the same approach used for the inverter. The schematic of the PLS NAND gate is shown, in Fig. 8.13, together with its transfer characteristic measured keeping the B input to V_{DD}.

Ring Oscillators

In order to evaluate the dynamic performance of the PLS inverter, different ring oscillators were designed. Figure. 8.14 shows a picture of a 9-stage ring oscillator exploiting PLS inverters plus an output inverter and a buffer helpful to drive the measurement setup.

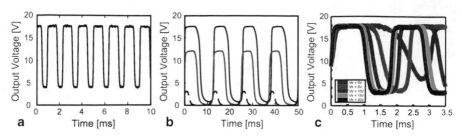

Fig. 8.15 Measured output of different ring oscillators: **a** a 15-stage ring oscillator, **b** a ring oscillator supplied by different voltages $V_{DD} = 20$, 10, 8 V (after 6 months shelf-life), and **c** a ring oscillator biased by different control voltages V_{tr} (inset)

The plot of Fig. 8.15a shows the time behavior of a 15-stage ring oscillator output. The measured inverter delay was $T_d \sim 45$ μs which is comparable with the fastest Zero-Vgs inverter reported in [9]: indeed in both cases the delay is dominated by the weak pull-down action of the output stage.

PLS inverters can add more functionality to a simple ring oscillator. Indeed, the presence of a tunable trip-point guarantees the functionality of the circuit even with reduced supply voltages. Fig. 8.15b shows the plot of the output voltage of a ring oscillator measured at different supply voltages, between 8 and 20 V. Moreover, the tunable trip point allows to implement a VCO [10]. In Fig. 8.15c the output frequency is varied from 300 to 560 Hz tuning the control voltage V_{tr}.

The three measurements shown in Fig. 8.15 reveal a different average delay for the inverters in the ring oscillator. For instance, the average inverter delay, observed during the measurement of the ring oscillator in Fig. 8.15b, was about 660 μs. This delay is more than one order of magnitude larger than the one observed in Fig. 8.15a. Such a big difference in speed is not due to the process variability, but depends on the ageing of the semiconductor. Indeed the ring oscillator of Fig. 8.15a was measured right after its realization, while the one in Fig. 8.15b was measured after the samples were kept for more than 6 months on a shelf. Aging was also confirmed by mobility measurements on test devices: after a few months mobility was found to be more than two orders of magnitude smaller.

Data Flip Flop

The Data Flip Flop (DFF) is an important building block since it can be used as a storage element in sequential logic, it can be employed as a delay element in digital filters, it can be used to synchronize asynchronous data, etc.

In our case, the DFF is used to build a 240-stage shift register. Our synchronous data flip flop includes 6 NAND gates and one inverter, for a total of 52 OTFTs using the PLS style (Fig. 8.16).

Fig. 8.16 Schematic of a synchronous DFF. (© 2014 IEEE. Reprinted, with permission, from D. Raiteri et al. [1])

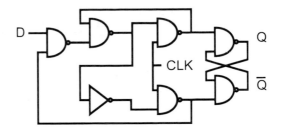

Shift Register

In order to demonstrate the actual improvement of the yield with a large functional circuit, a 240-stage shift register was designed using the PLS logic style. This circuit employs 13,440 p-type OTFTs achieving (to our knowledge) the largest transistor count ever demonstrated in a functional organic circuit. Our circuit features almost four times more p-type OTFTs than the microprocessor discussed in [11], which has been fabricated in the same technology.

The photographs of the full 240-stage shift register and of a 10-stage module are shown in Fig. 8.17a, b respectively. The shift register occupies an area of about 26×26 mm including 384 pads to allow the measurement of the output of each stage. One additional output buffer is included at the output of each DFF in order to avoid excessive load due to the measurement setup. The shift register is divided in 4 blocks, each containing six 10-stage modules.

Figure 8.18a represents the input (red) and output (green) of one 60-stage shift register block at 70 Hz. On the other hand, Fig. 8.18b shows the measurement at 30 Hz of all four blocks building the 240-stage shift register. Due to the presence of a buffer, the output swings between 7 and 17.5 V. For the first measurement, the trip point control was set at $V_{tr} = 16.5$ V and a digital input pulse was provided every 100 periods of a 70 Hz clock.

Fig. 8.17 Photographs **a** of the 240-stage shift register and **b** of the 10-DFF module. (© 2014 IEEE. Reprinted, with permission, from D. Raiteri et al. [1])

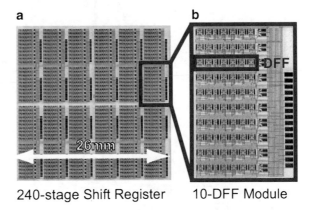

240-stage Shift Register   10-DFF Module

Fig. 8.18 a Measured output
after 60 stages at 70 Hz. **b**
Measurements of all the four
blocks building the shift
register. The four plots show
the output signals respectively of the 60th, 120th,
180th, 240th stages. In both
cases, the reduced swing of
the output data is due to the
buffer included to drive the
measurement setup

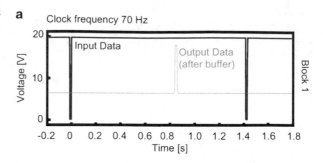

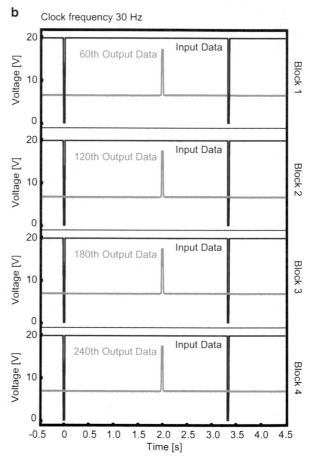

The control voltage V_{tr} was then varied to investigate the voltage range and the
accuracy that would be required by a control circuit able to provide automatic tuning of the trip point for maximum yield. For the measured foil, the full shift register
was functional, at 30 Hz, for V_{tr} changing from 14 to 20 V. For instance the four
plots in Fig. 8.18b show the input and the output signals for all the four 60-stage

blocks of the same 240-stage shift register when the bias is $V_{tr} = 18$ V and the clock runs at 30 Hz frequency.

The current drained from the 240-stage shift register supply is almost independent of the control V_{tr} applied to the shift register and amounts to about 400 µA.

8.2   Conclusions

Digital circuits are always present in integrated circuits either to implement digital functions or to serve other purposes. In this chapter, was discussed the design of a digital logic style suitable for three metal layers unipolar TFT technologies.

The proposed logic style achieves outstanding results with respect to the most common problems shown by large-area technologies manufactured at low temperatures. Indeed, the general aims of high gain and symmetric transfer characteristic are very difficult to reach when using these technologies even disregarding process variations and aging effects; and providing all the logic gates required by a functional circuit with a sufficient noise margin easily becomes a prohibitive task in actual prototypes.

The proposed logic style achieves the largest gain reported in literature using TFTs on film, exploiting positive feedback to overcome the low intrinsic gain of organic transistors. Moreover, to cope with the strong process variations that characterize TFT technologies on foil, a control voltage is included to tune the static characteristic of the logic gates. We firmly believe that no logic style will allow a widespread use of smart circuits on plastic foils, if such kind of control will not be available. Moreover, the needed control voltage is kept, with the proposed PLS logic style, within the supply rails.

The statistical behavior of 288 PLS inverters was compared to the one of the same number of Zero-Vgs inverters manufactured in the same area. Using the control voltage, it was possible to measure a noise margin larger than 5 V for all the PLS inverters under test (for $V_{DD} = 20$ V). Aiming at a soft yield of 90%, the better noise margin leads to an improvement in the achievable circuit complexity, which goes from 200 TFTs when using Zero-Vgs inverters to about 24 million TFTs when using PLS inverters (and adjusting the tuning voltage per group of inverters on a 4×26 mm^2 area).

In order to prove the actual yield improvement in a functional circuit, a 240-stage shift register was also designed. The fully functional circuit exploits over 13,440 TFTs, which is the highest complexity ever reached in organic electronics on foil to our knowledge.

These remarkable results to our opinion can pave the way to a widespread use of large-area electronics in more and more complex applications.

References

1. D. Raiteri, P.v. Lieshout, A.v. Roermund, E. Cantatore, Positive-feedback level shifter logic for large-area electronics. IEEE J. Solid-State Circuits **49**(2), 524–535 (2014)
2. H. Sirringhaus, Reliability of organic field-effect transistors. Adv. Mater. **21**(38), 3859–3873 (2009)
3. T. Yokota et al., Sheet-type flexible organic active matrix amplifier system using pseudo-CMOS circuits with floating-gate structure. IEEE Trans. Electron Devices **59**(12), 3434–3441 (2012)
4. M. Alioto, L. Pancioni, S. Rocchi, V. Vignoli, Exploiting hysteresys in MCML circuits. IEEE Trans. Circuits Syst. II Expr Briefs **53**(11), 1170–1174 (2006)
5. M. Alioto, L. Pancioni, S. Rocchi, V. Vignoli, Analysis and design of MCML gates with hysteresis, in *IEEE International Symposium on Circuits and Systems*, pp. 21–24, 2006
6. J. Rabaey, A. Chandrakasan, B. Nikolic, *Digital Integrated Circuits: A Design Prospective*, 2nd edn. (Prentice-Hall, Upper saddle River, NJ, 2003)
7. S.D. Vusser, J. Genoe, P. Heremans, Influence of transistor parameters on the noise margin of organic digital circuits. IEEE Trans. Electron Devices **53**(4), 601–610 (2006)
8. M.G. Buhler, T.W. Griswold, The statistical characterization of CMOS inverters using noise margins. J. Electrochem. Soc. **257**, 391–392 (1983)
9. K. Myny et al., Unipolar organic transistor circuits made robust by dual-gate technology. IEEE J. Solid-State Circuits **46**(11), 1223–1230 (2011)
10. D. Raiteri, F. Torricelli, A.H.M.V. Roermund, E. Cantatore, Design of a voltage-controlled oscillator based on organic TFTs, in *International Conference on Organic Electronics*, 2012
11. K. Myny et al., An 8-bit, 40-instructions-per-second organic microprocessor on plastic foil. IEEE J. Solid-State Circuits **47**(1), 284–291 (2012)

Chapter 9
Conclusions

This book illustrates innovative contributions to the design of electronic circuits for large-area and low-cost applications. This sort of applications is enabled by a cheap, low temperature processes to manufacture electronics on flexible substrates. Unfortunately this kind of processes only offers TFTs with electrical performance far below the silicon standards, and enables only poor uniformity and reliability. For these reasons, state-of-the-art circuit designs based on these technologies are still at their infancy and only first steps have been moved towards of the implementation of complex analog functions, the accurate conversion of analog signals to the digital domain, and the creation of large digital circuits with satisfactory yield.

With this work, we demonstrate that all these important aspects can be substantially improved while exploiting the same technologies used so far. Technology-aware solutions have been found facing the technology issues from the architectural, circuital, layout and device points of view.

Concerning analog circuit functions the following conclusions can be drawn:

1. Circuit simplicity should be pursued to avoid strict bias constraints, and to achieve sufficient yield despite the large process parameters variability.
2. Exploiting suitable topologies, circuit performance should be disentangled as much as possible from absolute parameter values, as it has been done in this work for the gain of the tunable transconductor and of the parametric amplifier.
3. An additional control can be a simple but very effective solution to tune some key performance to the wanted target in spite of process variations, e.g. the cut-off frequency in a $G_m C$ filter presented in Chap. 6.

In order to design circuits for data conversion with better linearity:

1. Careful layout in DACs, to cancel where possible gradient effects, should be pursued. With respect to matching properties, increasing the unity cell area must be done carefully, as it can soon be detrimental due to high probability of hard faults typical of these processes.
2. A rational choice should prefer converter architectures where the amplifier and comparator offsets do not affect the linearity, as successive approximation register (SAR) and integrating ADCs.

© Springer International Publishing Switzerland 2015   129
D. Raiteri et al., *Circuit Design on Plastic Foils,* Analog Circuits and Signal Processing,
DOI 10.1007/978-3-319-11427-9_9

3. ADC topologies based on matching should be avoided, since matching is obviously one of the weakest points of large-area electronics. In case matching is exploited, rely on matching of passives (e.g. resistors, capacitors) or exploit TFTs manufactured with amorphous semiconductors like GIZO. TFTs based on nano-crystalline semiconductors like pentacene should be avoided as unit elements in a DAC.
4. Integrating ADC architectures can drastically improve linearity and, at the same time, reduce area and power consumption, especially when integration is performed in the digital domain. Compact area is beneficial also for resilience to hard/soft faults.

Digital circuits play a very important role in any integrated circuits, and the same will be the case for the large-area low-cost applications we target in this work. About digital circuits:

1. Noise margin is the key figure of merit to be improved. Wafer-to-wafer variability, mismatch, bias stress, aging are not fully characterized yet, but still represent the major hurdle for (digital) circuit design. Noise margin should be robust against all these factors.
2. A symmetric input-output characteristic is mandatory to achieve acceptable noise margin, and a trip-point control can make the difference with respect to the yield of a logic style. An effective trip-point control using voltages within the power supply can allow self-calibration solutions.
3. Positive feedback has been shown as a viable way to provide a robust logic style improving voltage gain.

Following these guidelines, relatively complex circuits can be designed with an acceptable confidence on robustness, even if quantitative data on matching and statistical process variations are not provided along with the technology.

Also investigations around innovative device designs can result in better performance. To this extent, a parametric capacitor was designed to be used in a discrete-time amplifier, which provides a voltage gain comparable to continuous-time solutions in the same technologies, but achieves faster response.

This work proves the feasibility of complex circuit designs with large-area electronics and shows the sustainability of unipolar technologies (especially if featuring double-gate) against complementary ones in the prospective of addressing low-cost applications.

This boy had been unaware that he always expected his own particular answer to questions. The rest of the group helped him to see how his habit was irritating friends and foes alike. Very often quirks of this type (and other snags that interfere with a free exchange of ideas) can be brought out in a frank discussion following the playing of the role.

"So roleplaying's nothing new," you may say, "I do it all the time." Even if you do I would like to encourage you to try this method and these roles. As the conductor, your position is something like that of a football coach who sends his player out into the field with a pep talk. You support each player in his role while stimulating a frank conflict between players.

Roleplaying, as set forth in this book, helps you to dig out problems caused by the lack of listening and understanding. It can bring to the surface frustrations, guilts, and resentments. Roles are structured to help with the numerous problems of teenagers, all the way from how to give (or not to give!) advice and take criticism to how to give and accept love. Although help is given with each stage of the roleplay, the roles allow for and encourage the reader's own creativeness.

In facing today's world, the three "L's" of Listening, Leveling, and Loving are as important as the three "R's" of Reading, 'Riting, and 'Rithmetic. The need for the first two of these "L's," sensitive listening and leveling, is made drastically clear in roleplaying, for without them a roleplay will not move forward. Loving, the third "L," is important because the world the teenager aims for should be a world peopled with intimate friends who care, not just a world of mere acquaintances. Courtship relationships are brought out in a number of roles, from "puppy love" to manipulating someone into marriage.

Games are used to focus some of the main points brought out in the roleplaying. These games are a unique and useful feature because they provide the teenager with understanding and with a new way to react. The games "Mimic Charades" and "I-Thou Feedback," for example, teach listening techniques. Another game, "Manipulative Basketball," has players use exaggeration to understand and thus overcome manipulation. (Manipulating is using a non-leveling approach in relationships, which tends to be self-defeating.) The "Who Is Me?" and "PAC Pinball" games start a youth out on the adventure of discovering himself. The Symbol Game helps him become aware of his potentialities while encouraging him to set goals and become involved in something he is "turned on" by. One game helps youths give "Fuzzies" (compliments) rather than "pricklies" (put-downs). Another gives one player a chance to show love in adverse circumstances while allowing another player a chance to release pent-up resentment.

The roles in this book are simple and designed especially for the beginning roleplaying conductor—one who may, or may not, have had some counseling experience. A psychoanalyst or experienced therapist (who has the know-how to deal with difficult problems if they arise) can easily change or deepen these roles

9

to deal with more traumatic experiences. The book is not addressed to the psychiatrists or psychologists who are usually working with the abnormal on a one-to-one basis, for the roles try merely to deal with the communicating difficulties of normal youth.

A beginning conductor (that is, the director of the roleplay) will find it helpful to read over all of the instructions in chapter 2 and some of the suggestions with the roles before starting to roleplay. Some of this material, particularly in chapter 2, may seem quite technical to a newcomer. However, this book deals with many ways to get the most out of roleplaying, so don't be concerned to learn every technique or concept at once. After you have tried conducting for a while, refer back to chapter 2 for new ideas to try.

Conducting roleplays does not require any special wisdom or professional training; the skills you need are mainly those of listening and observing. As you become more proficient in this role of observer, you will be able to detect when one player is manipulating another or where the "child" in the youth has gained the upper hand. At first it would be well just to concentrate on giving your players the information they will need. You could start out just by reading the information to them (being careful to leave out material not meant to be given); but you will get much better results if you talk directly to your players, in a personal tone and using your own words.

Teenagers today are on the threshold of a dramatically changing world and need all the help they can get before being thrust into it. This book provides tools for you to help them. Dare you pass up the chance?

2

instructions

General Information

A roleplay, like an ordinary play, has a situation with a conflict. In roleplaying, however, a potential conflict is assigned to each of the players in a given situation, but they must supply their own dialogue and attempt their own solution. The solution in itself is not a primary aim in roleplaying. What is important is learning more effective methods of handling problems. In this book most of the roleplays concentrate on how the players handle the situation rather than on the issue itself.

You will have no need for scenery or props. A small room, preferably carpeted, will create the right atmosphere. Enough chairs should be supplied for all to sit comfortably around a center area that is used for the stage. Any person with authority can direct the roleplays in this book. Although the term "facilitator" is used in therapeutic roleplay, here the director will simply be known as the "conductor."

The conductor must remember that the main purpose is to help the role-players and audience gain new understanding. Roleplaying is an inductive method of learning and the insight is gained either during or after the playing. Keep in mind that in roleplaying the person usually acts, thinks, and feels as he

11

does in real life, but he is not under pressure so he can operate in a condition devoid of threat. It is because of this that the defects which cripple communication can be dealt with in a gentle way.

Roleplaying is a living, changing instrument and each little difference may be pointing out some important fact. Hurts, frustrations, and the emotions of anger and resentment should be watched for during the roleplay. Bad listening or speaking habits and irritating quirks of expression may be observed and helped during discussions. The way facts are presented should be noted and dealt with in a constructive way.

Emotion can be released in roleplaying. Teenagers often have a desperate need to rid themselves of feelings held in. In roleplaying they can do this in a fairly impersonal way. Most families cannot afford a psychiatrist just for dialoguing disabilities. Nor can they bring themselves to bother a minister or teacher for what might seem just minor trouble. Yet a little patience may ease hurt or tension, and may even head a youth off from the path towards serious crises such as violence or suicide.

A few general methods used in therapeutic roleplay can help point out snags in our dialoguing abilities. One method, which helps both the players and the audience to understand what is happening, is to stop the roleplay and ask the players how they are feeling.[1] Various answers are given, such as: frustrated, hurt, angry, puzzled, etc. These are often an indication of some dialoguing difficulty.

There are several ways to go about explaining such feelings:

a. If you are a beginning conductor you will probably feel more comfortable having a *discussion* about why players feel as they do, and how the dialogue could be handled in a better way. You may want to try the roleplay with two casts and then discuss why one player was getting himself and his ideas across better than another. Guide the players by your own insight into the situation.

b. If you are a more experienced conductor or someone who has used some counseling techniques, you might try a different method, namely "modeling." In simple language, this just means providing a model for your players to follow. If you can understand why a player is unable to get his ideas across, or why he is antagonizing the other players, or even why he is losing out in the conflict, you can get your idea over better by showing him rather than just telling him. The audience can also learn by this.

Basic Instructions:

1. Roles should not be attempted without someone in authority conducting them. This is especially important with teenagers. It is also advisable to forbid anyone to be a part of the audience who is not willing to take part in the roleplaying. This is to prevent a non-participating audience from making fun of the players.

2. In selecting a first role, a beginning conductor is advised not to try the

personification roles or roles without conflict for they are harder to conduct and harder to play. A role with a strong conflict and players who have a great deal of self-confidence are easier to conduct.

3. The conductor need not be concerned that teenagers are playing the parts of parents, grandparents, or small children. Roleplaying seems to have some magic of its own that surpasses even other drama. Once fully in the role, a person will play it for real: a teenage girl will suddenly *become* a grandmother or an eight-year-old sister.

4. To make roleplaying more effective, the conductor should describe the role to each player, individually, in front of the audience. The other characters should be out of the room so that they will only be aware of their own situations. (In the case of two casts, the two people with the same role should be briefed together. Also, the second cast should be out of the room when the first cast performs.)

5. The players will find it easier to use the actual first name of their co-players in the roleplay rather than the one given in the role. The conductor can use the terms "girl-part" or "Basil-part" or something like "father," "grand-mother," when he is giving his instructions, to help with this style of identification. If it seems difficult at first, the conductor should be guided by what seems comfortable.

6. In giving instructions, the conductor should try to be in sympathy with the player he is addressing, and also with the instructions he is giving. He should have a heart-to-heart talk with the player, giving him the facts, but not telling him how to use them or what to say. The conductor will be trying to intensify or heighten the conflict, and psychologically trying to encourage the player and put him in the mood of the role. If the conductor himself can get in the mood of the role, he may then subconsciously give it to the player.

Except in a few specified instances (see instructions for some of the games), the conductor should give the directions from this book *in his own words*, personalizing them by using the second person. With this personal touch, his sympathy for a role and for the person playing it will come across to all his players.

7. The role-dialogue is not meant to be given to the players. The aim is to allow the players to project their own personalities into the roles and nothing should hinder this. The role-dialogue given with each role isn't meant to be real dialogue but is given simply to show the conductor different points that will be helpful. Some of the dialogue is stiff and unreal, because I have concentrated on points rather than realism. At various times it will indicate: the type of player needed for the role, the aim of the roleplay, how leveling brings the conflict into the open and accelerates the action, how put-downs, sarcasm, and general nastiness slow the action, and how words indicate whether the parent, adult, or child is predominating in the player.

8. The conductor should be careful to set the scene so that each player is

plunged into the role and can feel a part of it. The opening scenes given are simply one way the roles can be started. If some other scene has more meaning for your particular youths, by all means use it instead.

9. The players can bring in additional complications along the lines of those in the role. They should always feel free to pursue complications that have more meaning for them. (Sometimes the problems that youths find hard to express surface in this way.)

10. Remember, there is no set way that a role has to be played. Each time it will differ because the people involved are different. Changing just one player could make a whole new ball game. In other words *do not force a solution*. No one should be made to feel that he is being pushed into a mold. The solution must come from the players or from a suggestive "modeling" by the conductor. It should aim at being a no-win, no-lose compromise. Sometimes it may just come to a draw where one or the other of the players is only willing to think it over. Remember, *how* the player reaches the solution (or aims towards it) is the important thing.

11. The conductor can give out information which one person is withholding, and he can add more information which would help bring about a solution. This can be accomplished by a faked phone call, or by sending in another player with specific instructions. A note can be used when only one player should have the added information. Most technical problems can be solved by giving new information in the same way.

12. A time-shift can be used if it will help in solving a problem or reaching a solution. The conductor simply says something like: "It is now the evening of the next day, and . . . ."

13. Sometimes a player is using the role only as an emotional release and his mind is not set on problem solving. The youth may need to rid himself of some suppressed feelings so this should be allowed to continue for a short time. However, it should not be allowed to continue long enough to bore the rest of the cast and the audience. The conductor should call "cut" and use his own judgment in dealing with it. A simple way is to mention that the roleplay has gotten away from the problem and suggest that the players get back to the issues of the role.

14. A question a new conductor might ask is: When do I stop the roleplaying? Most conductors quickly learn when to stop, but if you are just beginning, a good rule to follow is to watch for the interest of the players and the audience. Some roleplays may be long and drawn out. When something constructive is taking place, however, it will be dramatically interesting even through minutes of silence. If a conductor is in doubt, he should simply stop the playing, get the reaction of the players up to that point, and ask them if they would like to continue. Sometimes the roleplayers are having a great time and would continue indefinitely. The audience should then have some say in its continuance. When a compromise or a solution is reached the roleplaying should

be stopped. Usually when problem solving is not taking place (and it seems impossible to achieve it) the roleplaying should be discontinued.

15. After the roleplaying, be sure to allow the players to give their reactions before opening up discussion to the audience.

Selection of Players:

Players will sometimes choose their roles voluntarily. This in itself can bring out a number of interesting insights. However, a great deal more can be accomplished if the conductor puts players into roles where he can see they will he helped. It stands to reason that the better the conductor knows his players (and the more sensitive he is to their needs) the more effective the roleplaying will be.

In selecting players for most roles, you will probably want them evenly matched, or as close to this as you can come.

a. The first exception to this would be when there is no conflict, so that the abilities of the players are not important.

b. The second exception would be where identification with the role is more important than the ability of the player.

c. A third exception would be for a new conductor starting out. Two extremely shy youths might be evenly matched, but in a role without a conflict neither might want to take the initiative, making an inexperienced conductor at least uneasy.

d. The fourth exception would be certain roles that are designed to point out the comparative abilities of the players, and players should be selected for them accordingly. In this book, a person with a strong personality will be designated as a *heavyweight* in contrast to the shy person designated as a *lightweight*.[2] Roleplays that require heavyweights and lightweights in opposition are ones like "Unwarranted Help" (p. 27) and "Did It Have to Be Divorce?" (p. 114).

These ability roles are given for a twofold purpose:

First, the conductor should be trying to show a person who has strong opinions (and the confidence to back them up) that sometimes he may have to tone them down, so that another person who does not come on as strong can be allowed the freedom to express himself and respond. It might help to point out to the heavyweight player that to win is not the only consideration in interpersonal relations, and that other players have needs that he ought to be aware of. The authors of *The Intimate Enemy* suggest that heavyweights should, figuratively speaking, hold one arm behind their backs when engaging a lightweight player.

Second, the conductor should be trying to help a lightweight player speak up for his rights. If possible an attempt should be made to show the lightweight player why he is failing to gain his points. Respect for oneself and for one's rights is a quality everyone should have, and it should be worth fighting for. Leveling about these feelings is a first step to winning. In *The Intimate Enemy,*

15

Bach and Wyden use the lack of leveling to head the list of drawbacks to the success of their methods.[3]

e. A fifth exception would be similar to the one above, but where a certain situation is needed and a player with more fighting ability would push harder; for example: "Broken Plans for Marriage" (p. 109).

What to Watch for and Bring Up During the Discussion:

1. A major stumbling block to any dialogue is lack of leveling. The same applies in roleplaying. When players hold back their own feelings, or the facts they have been given, the roleplay becomes bogged down with nothing moving. The conductor should call "cut" and ask the players how they are feeling. Various answers—from feeling frustrated to feeling confused—will indicate the plight of the players. The conductor could then "model." He simply steps in and plays one of the roles—telling the truth in a simple, but gentle and loving way. It gets amazing results!

2. Sometimes, during a crisis or immediately afterwards while the situation is still sensitive, someone will start preaching or giving advice. If this happens during a roleplay, point out that this tends to antagonize instead of resolving difficulties. Preaching, advice, and judgmental suggestions almost always alienate people. In roleplaying it may provoke a sharp retort. Thomas Gordon, in his book *Parent Effectiveness Training* (hereafter called *P.E.T.*), has listed preachiness among the responses he advises people not to make if they want to engage in "Active Listening." These are listed in this book with the game, "I-Thou Feedback" at the end of chapter 3 (p. 39).

3. Like a remedial teacher, you should look for weaknesses in the way a player presents his facts. A player who has a poor opinion of himself may be defeated before he even starts. Watch for the player who presents his facts in a halfhearted way as if he did not expect to win. You might use some modeling here to advantage. Present the facts with firmness and conviction and demonstrate the difference it makes on those who are listening.

4. Each player should have the freedom to make his own decisions. Watch for places where this is being interfered with. This does not mean, however, that players who are taking the parts of someone in a position of authority cannot warn others of the dangers or consequences of their decisions.

5. Watch for the places where a player is not listening carefully enough. He may be so wrapped up in his own view that the other player is not coming through to him. This hinders the solving of the conflict. Watch also for the player who is so emotionally involved or is so impatient that he is closing off his listening capactiy.

6. Watch for parts of the instruction which a player forgets. There is usually a reason for forgetting, and it may be important. Ask the player if he wants to take a closer look at why he forgot. (In a few roles ignoring facts may be

prudent; for example: "Forces Crippling the Ego" [p. 77].)

7. Also, watch for the times when players distort or block out information. The player may have a strong opinion, prejudice, or bias, and this may influence how he hears and deals with the facts.

8. Watch for "body language." Tone of voice, gestures and other body movements, expressions on the face, and just the way a word is spoken can sometimes take on more importance than the word itself. A player's demeanor may be unconsciously shouting something completely opposed to his words. Even a word like "oaf" can be a term of endearment if handled in a certain way.

Very often, youths are unaware of how they are coming across. Though sometimes in roleplaying the players can understand their effect on others by the results they get, there are other times when they are completely unaware of what caused such a reaction. The conductor should jot down these times during the playing of the role. Then, during the discussion, he can jog the memory of the players by bringing up the instances of the subconscious tone of voice and the reaction it received.

9. Sometimes it is better to let two people work out a conflict alone without a third party's interference. If someone does interfere but not in a way that significantly disrupts the playing, let it go on and point it out later in the discussion following the roleplay. (Disruptions are sometimes inserted as part of the role by using a prop player—one with definite instructions to follow until the scene becomes emotionally difficult. This should be brought out in the discussion period so all are aware of the reason for the disruptions.)

10. Occasionally one player may be hindering the dialogue and may have to be dealt with before the action can go forward. It is important that he be dealt with in a way that helps rather than antagonizes him. The roleplay can be stopped and the players asked how they are feeling. This may bring out the reason for the offending player's actions or it may bring out the other players' reaction to a dialoguing quirk. If, in a roleplay with three characters, it is established that antagonism between two of them prevents them from dealing with the third (as in RP 10, p. 83), the third player may be advised to deal with each of these antagonists individually.

11. Watch for parallel speaking. Players who are talking about the same subject but are developing their own cases so independently that they do not intersect. In other words, neither party is listening to his opponent.

12. We can think much faster than a speaker can talk.[4] How are the players using this extra listening time? They can use it to best advantage if they analyze their opponent's points and the direction he seems to be going. They should, however, listen not merely to his intellectual points but also to the direction of his feelings.

13. Salesmanship is great for selling but not always for solving interpersonal relations. Watch for the person coming on too strong or using propaganda tricks.

Nichols and Stevens, in their book *Are You Listening?*, list seven propaganda tricks that salesmen use.[5] Here are some ways these tricks crop up in roleplaying:

a. *Name calling*—using labels that produce emotion, usually with an implied threat. A roleplayer could say: "Your point is distasteful to anyone of intelligence." "You sound like a Communist."

b. *Glowing terms*—glittering generalities that put our own cause in a favorable light. Example (taken from "Broken Plans for Marriage" RP 14, p. 109): "Traveling with the Ballet all over the world will make a full and exciting life which I'm sure you wouldn't want to miss. We'll never lack for money. Friends, hobbies, fame—we'll have a wonderful time together."

c. *"Authorities say"*—making our cause respectable by association. Example: "I know you have a lot of respect for science and this is what scientists have found."

d. *Testimonials*—using respect for well-known people to promote our own point. A roleplayer could say: "Lincoln used this point and what's good enough for old Abe is good enough for me!"

e. *Editing*—card stacking material to promote our own cause, neglecting to mention the other side. Example (taken from RP 6, p. 49): "Constance honey, this income property leaves me no time for my inventing. There's always a drippy faucet, or a door lock that needs fixing, or a tenant who's lost his keys, or . . . ."

f. *Using Trust*—playing on people's trust of those like themselves and their distrust of others. A roleplayer could say: "I know you can understand my point—we're both men and we know how a man feels when he gets upstaged by a woman."

g. *Bandwagon*—playing on follow-the-crowd instincts. Example (from RP 5, p. 46): "Gosh, Lee, everyone will be there. Do you want them to think you weren't invited?"

14. Watch for *I-Messages* and *You-Messages*.[6]

You-Messages tend to be a put-down and do not help a relationship. Examples: "Will you stop that? You're acting like a child!" "You're old enough to know better than that!" "You're so messy and disorganized! Why can't you do things right for once?"

I-Messages are a form of leveling, telling how *you* feel about a situation rather than what you feel someone else should or should not be doing. This type of message is more acceptable to another person. Example: "I'm discouraged when I walk in and see this mess. A hang-up of mine is trying to get a meal in a dirty kitchen."

A teenager to a younger sister, instead of saying, "You're a stinking little pest, and you bug me!" might try, this: "Susy, I'm upset! I'm trying to get my homework done and I can't do it with you bothering me. Let me get it done and we'll play afterwards."

Some tips about I-Messages from *P.E.T.* by Thomas Gordon that can help a conductor with discussion after a roleplay:[7]

a. Watch for what seems to be an I-Message, but really is a You-Message. Example: "I get upset whenever *you* goof off."

b. Encourage players to accentuate the positive, not the negative in their I-Messages.

c. An I-Message will not work if the player understates how he feels. Encourage players to express emotion in the degree they are feeling it.

d. Watch for "I am angry" remarks. They are usually You-Messages for they are directed at someone else and are only a secondary feeling. A player's own embarrassment, hurt, disappointment, resentment, or needs may trigger an angry feeling, and he may direct it at an opponent in the roleplay. A blameless player may wonder what he has done to cause this.

15. Watch for the player who tries to take on the problems of the other players. He will tend to do this in real life. Like Atlas he is trying to carry the whole world on his shoulders. Both growth and dialogue are promoted in a mutual-rights environment. Every person has needs and rights, and when his needs are going unmet or his rights are abridged, he is said to have or to "own" the problem.[8] And it is up to *him* to find a solution. Sometimes one player will try to "own" another player's problem (this is scripted in RP 16, p. 114)—and this will prevent the real owner from coming to terms with it. With help, the roleplayers will learn to distinguish whose problem it is. When it is their problem, they will shoulder it. When it is not, they will recognize this and possibly learn better ways to help.

16. Watch for the player who does not finish a thought or a sentence before rushing on to his next idea. This quirk is confusing and a listener's reaction may be sharp or irritated.

17. Watch for the player (or member of the audience) with irritating habits that distract from the roleplaying. Some people have the habit of tapping a pencil or their feet, others of drumming their fingers. Still others keep watching the clock or glancing at their watch. Usually these actions indicate to the others that this person is bored. Such actions also indicate that the offending person is not bothering to listen very carefully. This is impolite if not rude. During the discussion find out if the offending person has been listening and ask the others how they feel about this inattention.

18. Watch for the player who has the bad habit of breaking in and finishing a sentence when he thinks he knows what the other person is going to say. He may be right, but right or wrong it is distracting to the speaker.

19. Watch for players who speak in a monotone, without expression or seemingly without energy. Are they finding it hard to gain their point? Modeling can be used here to advantage, to show how expression in a person's voice and some aliveness can get better results.

20. Words are only the stimulus or the symbol of the meaning underneath. Players should try to understand what another player is trying to say. Watch for the player who is taking the words literally, and makes no attempt to search for a meaning. He seems unable to "read between the lines." Sometimes a player will even say, "I don't care what you meant. What you *said* was . . . ."

21. Be aware of players who ask questions but will not accept the answers they are given. Some keep on asking the same question until they get the answer they want. This habit can be irritating and a new more acceptable way should be learned. (Young people acquire this habit from parents or teachers who use the device frequently.)

22. Look for manipulation. Not just the aggressive kind, but the soft-sell type: "You go ahead, Dear, and don't worry about poor little me." Watch for the player who plays the role like a wounded hero for all the mileage he can get. This is a form of manipulation; so is humility.

Watch for places where a player is being pressured or manipulated into another way of thinking. How does he react? Does he handle it quickly and easily? Does he handle it by using another type of manipulation? This should be brought up in the discussion session after the roleplay.

A book by Everett Shostrom, *Man, The Manipulator,* gives names to the types of manipulators. The entire book is worth reading for a roleplaying conductor. Shostrom's eight manipulative types are listed with the game "Manipulative Basketball" in chapter 4 (p. 63). After the manipulative game there is a list of traits which are typical of manipulation, along with their opposites—qualities characteristic of what we call "actualizing."

23. Watch for the conversational novice who does not know where and how to become a part of conversation. He breaks in at the wrong time and is hurt by the lack of response. You can help this reticent kind of person by putting him into a roleplay with more than two players.

24. Sometimes an aggressive player, in his desire to get his own point across, will completely ignore what the player before him said. If one player has ignored another several times, the conductor might stop the roleplaying and ask the ignored player how he is feeling.

25. There is a parent, adult, and child in every individual. At different times one or the other of these personal parts gets the upper hand and expresses itself in words and actions. In the teenager this is more noticeable than at any other time in life, for teenagers slip easily from the child to the adult and then back to the child—or perhaps to the parent. In roleplaying the conductor will become increasingly aware of when one of these states dominates his players. Berne, Harris, and James and Jongeward have made a study of these times (when one of the personality characteristics is predominant) and call them *Ego States*. PAC Pinball, a game in this book, (p. 98) is based on these Ego States and the reader may wish to refer to it for further knowledge on the subject when he begins to notice these states in his conducting.

26. Allow players the right to refuse a role, or to quit a role if it becomes too uncomfortable.

27. It will be obvious to the conductor that these roleplays can be used many times with the same players; they simply change the parts they play. It may *not* be as obvious that a player who was having trouble with a particular part could repeat it sometime later pitted against different players. In this way they learn new ways of reacting and gain proficiency in them.

These general instructions are important because with roleplays you are *not* dealing with just particular roles, but with *people* who are *roleplaying.* After each role suggestions of what to watch for are given. These are based on what will probably happen. But bear in mind that they are only suggestions. Some suggestions will apply for one playing of the role; other suggestions for another playing of the role with different players. It is possible to conceive of circumstances where none would apply. There would be nothing wrong with your conducting or with the role. It is just that in roleplaying you are dealing with unique and unpredictable people. Every one of these people has different experiences and feelings. The roleplay is just the setting, the situation, the start.

Keeping the above paragraph in mind, if the roleplay goes off on a wild tangent, allow it to continue as long as it remains vital. When it becomes bogged down and either the audience or the players lose interest, stop the playing and find out what has happened. It can then be gently brought back to problem solving, or ended because of the impossibility of a solution.

Suggested reading:

Muriel James and Dorothy Jongeward, *Born to Win* (Reading, Mass.: Addison-Wesley Publishing Co., 1971); and

Everett L. Shostrom, *Man, the Manipulator* (New York: Bantam Books, Inc., 1968).

3

li/tening

Dialogue is often impossible simply because we do not listen to what the other person is trying to say.

A friend of mine, a social worker and a very understanding person, finds it hard to forget a cry for help that she simply did not hear.

Just after getting home from a trip, my friend, Kitty, went over to get a parakeet she had left with a neighbor. After the usual small talk the woman asked her in an offhand manner, "Kitty, have you got a little time to talk? I've got . . . a . . . a problem I'd like to discuss with you."

Kitty was rushed (there were so many things she had to do because of being away) so she said, "Gosh Martha, I can't stop now, but I'll call you early tomorrow morning. Okay?"

That afternoon the woman went into a closet with a shotgun and blew her head off. Kitty said that Martha was a quiet person who talked very little, and she should have known that it was important. Very few of us would have heard that cry, but there are voices of people all around us that hide their desperate need.

The roles in this chapter deal with the different types of sensitive listening, from attention to "body language" to the type of listening that tries to hear the reason and the emotion behind the words.

FEAR OF CRITICISM, OR POOR LISTENING?

Roleplay 1

*Basil: Basil has been barking at everybody for days, mostly because he hasn't seen Mae since their quarrel of a couple of days ago. He misses her a lot but is almost too angry to talk to her. She rather bluntly suggested that he talk over their problems with *her* instead of running to his mother. He's no mama's boy and she knows it! His mother is a grand old gal and a guy needs a sympathetic listener. Would she rather he go to some girl? He knows plenty of takers in that category! If she would just keep her big mouth shut and listen once in a while instead of interrupting with stupid suggestions that would have no chance of working out. Like the stupid things she said when he had that run-in with his boss at his after-school job. What does she think he is—a nincompoop? Maybe he ought to date that quiet girl in his chemistry class!

Mae: Realizing she has spent very little time talking to her boyfriend in a personal way, Mae now is determined to find out why, even if it means another quarrel. And they've had plenty of those lately! Why does he seem to shy away from talks on their relationship? This time she's not going to be put off by his authoritarian tone or his big grandiose explanations. And she's not going to let his anger or sulking in silence bother her. She wishes she hadn't brought up the subject of his mother in their quarrel but she had resented his bringing in a third person. She wishes she didn't love the dope. No phone call in two days. Is he calling it off?

[As the scene opens, it is nine at night and the doorbell rings. Mae goes to the door and is surprised to find Basil. She invites him in.]

Mae: You've been avoiding me lately, Basil. Why?

Basil: I've been talking to my mother instead.

Mae: I'm sorry about that, Basil. No one likes to have their quarrels aired with other people.

Basil: Mother sensed we had a quarrel. I was just hoping she could help me understand something. Lately when we're together, I always end up feeling angry.

Mae: Always? Maybe you do it on purpose so you won't have to continue a conversation with me!

*Role-descriptions and dialogue are for the conductor only; they are given as guides.

24

Suggestions for the Roleplay "Fear of Criticism, or Poor Listening?"

Selecting Your Players. This roleplay pictures a girl who talks instead of listening and a boy who is a little too sensitive about criticism. However, you don't need players who show these characteristics obviously, since everyone has them to some degree. Players should be of equal fighting ability or as near to this as possible.

Instructions for Players. Remember, the descriptions and the role-dialogue are for the conductor. You should be trying to instill confidence in your players by sympathizing with each one in turn. To Basil you might say: "Who does Mae think she is, making cracks about your being a mama's boy! Is she just trying to get you sore?" Etc. To Mae you might say: "You wonder what's wrong with Basil lately. You're listening to some of his troubles and trying to help him and suddenly he's sore about something! If you didn't love him so much you'd tell him you never wanted to see him again!"

Remember to set the scene for your players so that they can more easily be a part of the role. The scene given here is only a suggestion. If another scene would have more meaning to the particular players you're working with, by all means use it. To bring Basil into the role you should add something like: "You know it's late but you've decided you *must* talk with Mae. You have gone over to her house and have rung the bell. Looking through the window, you see that she, and not her mother, is coming to answer the door." Then set the scene for Mae by something like: "The doorbell has just rung and you wonder who would be ringing it so late. You go to the door and there stands Basil."

What to Watch for. Sensitive listening and leveling are important here. The role-dialogue given shows how leveling moves the players closer to the problem if not always to the solution. It shows how the action moves forward when one player is able to swallow a nasty remark or a sarcastic one. The role-dialogue is also intended to show how careless words slow the action and how exaggeration tends to bring forth anger. These are some of the things you should be watching for so that you can bring them up in the discussion, or model a better way.

In modeling you might try Thomas Gordon's technique of "active listening," which he describes in his book *P.E.T.*[1] This is the kind of listening that gives a feedback on what the listener is hearing. In giving feedback, you should try to express what you feel is the emotional impact behind what the sender is saying. Suppose, for example, that Basil is telling how his employer yelled at him to pick up the bag of potatoes or growled at him when he entered a room. Active listening might be expressed by saying: "I gather what you're saying, Basil, is that you didn't mind his orders, but it's the way he gave them to you that you object to and that makes you angry." This is not offering advice, but just showing the other person that you are listening and understanding.

Discussion Following the Roleplay. The discussion is important especially to those playing the roles. If two casts are used, there should be a discussion after each roleplay. Each person in the cast should be allowed to give his feelings before the roleplay is opened up for general discussion. Ask each player in turn how he was feeling during the roleplay and at the end. Encourage everyone to discuss any problems they encountered. Sensitive and active listening is the keynote of this role. You should explain active listening, then give the members of the casts (and then the audience) a chance to state where they think this type of listening could have been used to advantage. Leveling is always important and, if possible, you, as conductor, should point out where active or sensitive listening helped free the other person so that he could level.

Games and discussion questions on listening are given at the end of this chapter. The game, "I-Thou Feedback" (p. 38), deals especially with active listening.

UNWARRANTED HELP

Roleplay 2

*Carol: Because she has never been able to make her mother understand her point of view, Carol realizes that more and more she has been going along with her mother's ideas. Today, she has brought home, on approval, a darling ankle-length skirt like the ones her friends are wearing. Her mother has expressed disapproval of the long-length style for afternoon wear. It is very important to Carol that she keep the skirt. More than the item of clothing is involved here!

Eve: The mother of Carol. Eve is a dominant woman who does not seem to understand her daughter. To her, Carol seems to have such a hard time making up her mind. Even when she does make it up, you can never be sure she won't change it a number of times. One should make a decision and go with it. But she's basically a sensible girl, not addicted to doing things like wearing ankle-length skirts to school.

[Scene opens when Carol comes into the living room carrying a box containing the skirt.]

Carol: Can we talk a little before dinner, Mother?

Mother: Of course, Dear. What's in the box?

Carol: I'll tell you about that later. First of all, I want to remind you that I'm almost as old as you were when you got married. It's about time I started making my own decisions, even if they're wrong!

Mother: I'm not trying to make your decisions for you, Carol.

Carol: Maybe not consciously. Remember the talk we had about the hours I was to be in on weekends?

Mother: Yes. You made a decision and then changed your mind.

Carol: No, Mother. *You* made the decision. I just got talked into it. I'll grant that it's your prerogative to set hours, but you're wrong about the time!

Mother: But you agreed!

Carol: No! I couldn't make you understand, so I gave in.

Mother: Eleven o'clock is late enough to be out. Twelve-thirty is ridiculous!

Carol: Ridiculous to you—not to me! I have my friends to deal with.

*Role-descriptions and dialogue are for the conductor's use only. They are given as guides.

Suggestions for the Roleplay "Unwarranted Help"

Selecting Your Players. This is a role for lightweight and heavyweight players. A dominant girl should play the role of the mother; a shy personality the daughter. If you are using two casts and you have players that are very obviously lightweight and heavyweight players, you may want to concentrate on the heavyweight in one cast and the lightweight in the other. In this case you would then have them opposed by players of normal "weight." Since the second cast will be out of the room when the first cast performs, it would be best to allow the lightweight to be in the first cast so she can learn by watching how someone else plays her role.

Instructing Your Players. Remember you are *not* instructing your players in what they are to say or how they should feel, but in what they should know. In this case, they should *not* know they are playing the role of a lightweight or a heavyweight. If one asks, "How do I start?", you can counter by telling her to do what she normally would do. If that is by saying nothing, and just sitting there, allow her to do it, for that is probably what a shy person would do. It will be up to the other person to break the ice. If they both sit there and say nothing, allow it to remain that way for a time. Plays sometimes start in this way. Do not allow your audience to make fun of the players! This is important for teenagers. It is one of the main reasons why someone with authority should be conducting the roleplaying. Equally important is that your audience be composed *only* of those who will be roleplaying. To Carol, the daughter, after giving her some of the facts, you could say: "No matter what happens, you are going to keep that skirt. You have promised yourself this." To Eve, the mother, you can say: "Your daughter has such a hard time making a decision. And even then she doesn't stick with it! Why is it so hard for you to understand your daughter? You don't have this trouble with your son." Encourage your players to use other things besides the skirt that have personal meaning for them.

Remember, after giving your instructions you must set the scene for each of your players. It does not have to be complicated or long; just be sure to give them enough background to plunge them into the situation—so they feel a part of it. To Carol, the daughter, you could say something like: "You pick up the box with the skirt and go into the living room where your mother is sitting reading the paper." To Eve, the mother, it might be: "Your daughter has just entered the room with an odd expression on her face. She is carrying a box that looks as if it might contain clothes of some kind. You put down your paper wondering what the box contains."

What to Watch for. The dialogue given here (besides expanding the plot) shows the aim and direction of the roleplay. Both players should be using sensitive listening so that each can understand the other's point of view. Active listening—which goes beyond the words to the meaning and emotions behind them—can be used to advantage here. Are your players using any form of it? Do your players seem to be sensitive to each other's facial expressions and tone of voice?

Watch for defensive actions on the part of the lightweight player. How does she handle aggression? When backed against a wall, how does she react? The dialogue given here shows how the daughter can handle the situation firmly without resorting to any defensive actions such as put-downs or biting remarks. Does the lightweight player sound as if she believed in the facts? Does she present her facts in a low-key manner seemingly without energy? Watch for the heavyweight who comes on so strong that the lightweight is not allowed even breathing room.

Leveling is important here, for unless the lightweight tells how she feels and the heavyweight gives her reaction, arriving at a solution is difficult.

Discussion Following the Roleplay. The discussion will depend a great deal on how the role was played. Your dominant character may not have been as aggressive as you expected. She should then be told that she played the role with sensitiveness to the needs of the other person. If aggressive, she should be told (as was mentioned above in the general suggestions) that winning is not the most important item in interpersonal relations, and that she might try tying her hand behind her back, figuratively speaking. (This should be done tactfully, perhaps even speaking to the heavyweight privately, so as not to make the shy person uncomfortable.) If the lightweight player was unable to keep her vow of not giving up the skirt, the conductor might try helping her by modeling the role with her opponent.

DRUGS AND LISTENING

Roleplay 3

*Mike: A thirteen-year-old boy who has just been dismissed from school for smoking marihuana. He has been walking around trying to figure out a way to explain to his parents what happened. Will they understand that he needed to be a part of the gang? Could he get by with telling them it was just the first time? What if they found out about the pills?

Beth: A woman about thirty-five. She is horrified when the school phones and tells her that her son has been caught smoking marihuana. He was dismissed from school early but has not come home. She blames her husband for always being too busy to take an interest in the boy. Was smoking pot all that was involved?

Bart: Alarmed when his wife calls and tells him their son has been dismissed from school for smoking marihuana. On the way home he remembers when he sneaked a smoke behind the garage. But not marihuana! He is surprised that his son is old enough to smoke—or want to! He had been thinking of him as a baby. He realizes that he does not know his son.

[Mike walks into the living room to find both of his parents waiting for him. He can tell by their expressions that the school has called.]

Mike: I suppose the school has called?

Father: Yes, Son. Can you tell us about it?

Mike: Look, it's no big deal, Dad. I just had to show the guys I wasn't afraid. Wouldn't you know I'd get caught on my first attempt?

Father: I can see that it's important for you to be a part of the gang and be accepted by your peers, Mike.

Mike: Yeah, Dad. I felt kinda left out. Then, I . . . .

Mother: It would have been better for you to have been left out of *that* gang, entirely! I don't approve of some of the boys Mike calls his friends, Bart.

Mike: Who do you mean?

Mother: Guy and Bob. I understand from some of the P.T.A. mothers that they're using harder drugs. What have you got to say about that, young man?

*Role-descriptions and dialogue are for the conductor only; they are given as guides.

30

Suggestions for the Roleplay "Drugs and Listening"

Selecting Your Players. Your players' fighting abilities should be about equal, but the role can be played with any "weights." Try to compensate for a lightweight with more instructions. Any sex can play the boy part. The parent roles are not prop roles (i.e., simply used as foils) but are dealing with the Parent Ego State of the players (parent within the players).

Instructing Your Players. Remember, the descriptions and role-dialogue are only for the conductor. You will be addressing your players in the second person in your own words, and trying to give them confidence in the role they are to play. To the boy, Mike, you can say: "What are you worried about? This should be no big deal. Marihuana is certainly no worse than whiskey and both your parents drink. They couldn't have found the 'uppers' or that joint you sewed in your coat or they'd have said something about it, wouldn't they?" Remember to bring the boy-player into the role as you finish: "You see your Dad's car is home so they are probably waiting. Might as well get it over with! You walk in and can tell by their expressions that they know." To the mother, Beth, you could say something like: "It's Bart's fault! If he had taken more interest in the boy this wouldn't have happened. You've had a funny feeling for some time that something was wrong." Etc. To plunge her into the scene you can say: "Well, here's Mike and he does look worried. Is it more than marihuana?"

Remember, you must give certain facts, such as the boy's being dismissed from school for smoking marihuana, but each player should be given a little different slant with some other facts. To the father, Bart, you might say: "Your business has put such demands on your time that you haven't had a chance to know the boy. But boys will be boys. Isn't Beth making too much of the whole deal?" And to get him into the scene: "Here is Mike now and he does look worried. But who wouldn't be?—sent home in disgrace, getting bawled out by the principal and now about to get bawled out again."

What to Watch for. The start of the role-dialogue given here shows that no matter how gently the parents begin, the boy will be put on the defensive. Sensitive listening and gentle questioning are necessary here to allow the boy-player to be free to level. (Not only about the other drugs he is taking but about some of the deeper feelings the player might think caused it.) If the boy-player can be made to feel he is getting both understanding and love, he is more likely to want to level.

Gordon's "active listening" mentioned above (where the listener gives feedback on what he thinks the feeling is behind the sender's words) could be used to advantage here. The father in the role-dialogue here is showing this. Watch for the players who use some version of it, and others who talk instead of listening. Look for the players who sit and listen without giving any feedback on what they are hearing—not even a reaction. The player of the boy-part may interpret

31

this as anger or condemnation on the part of the parent-players. This will make him feel even more uneasy. Watch out also for the boy-player who feels that the sensitive listening should all be on the side of the parents and does not listen to what they are trying to say. Active listening should always be given a chance to go both ways. With a feeling of rebellion a player can be closing off the love and understanding he is getting from the other players.

Although the role-dialogue given here shows the father exercising caution and the mother coming on strong, watch for authoritarianism in either of the parent-players. Young people tend to parrot their parents in playing a parent role. This will take place not only in the roleplay but also in real life when the youth-player has children of his own. This is the parent input of the PAC (parent, adult, child) concept of Berne and Harris, mentioned above (see p. 20). Watch to see if both of the parent-players are going deep enough to discover all the facts.

Before the role is considered complete, some knowledge of the extra drugs should be brought out and the boy's reason for using them. If the boy refuses to tell about the capsule drugs, the conductor can fake a phone call from the principal of the school who says he has just heard it rumored that the boy has hard drugs sewed into the lining of his coat and the parents should check it out. The conductor could also send in another player (a prop player) who would volunteer the information.

Discussion Following the Roleplay. Sensitive listening and leveling are again the keynotes. The players should be asked how they were feeling and what caused them to respond the way they did. Bring up what you have observed of the parent input or the Parent Ego State of the parent-players. There can be some discussion on this, since the parent-players may not realize why they played the roles as they did.

Active listening should be explained. If any of the players has used a version of it, this should be brought out. Explain that this type of listening can show the other person you are trying hard to understand what he is telling you. If leveling has been a factor in the roleplay, bring out where it helped and where the lack of it slowed down the action. If an aggressive manipulation was used bring this out in the discussion.

The players and the audience may like to practice some of the listening games explained at the end of this chapter. After the games the young people may like to try another role that requires sensitive or active listening.

(Although the role is written primarily for teenagers, a conductor might also like to use it with *parents* who are having trouble with their teenagers. Have someone else besides their own child play the boy-role. It can be played by anyone of any age or sex. Watch for authoritarianism then, and for conflict between the parents that slows the action.)

THIS CRAZY WORLD

Roleplay 4

*Pricilla: A conservative girl of sixteen raised in a strict religious family, Pricilla felt out of place with her childhood girl friends because they were always talking about the pill and experiments with sex. She found the talk distasteful and said so. Because they seemed sex-mad to her, she had been spending a lot of time with a mannish girl who didn't date but was interesting to talk to and excellent at all sports. Her former friends started teasing her about having "gay" tendencies. Ignoring it at first, she began to wonder if they could be right, for she *was* repelled when a date made an awkward pass. Of the boys she dated she liked her chemistry partner the best; he was fun to be with and talked about other things besides sex. As her worry about being gay increased, she encouraged him and began to date him steadily. Her girl friends were delighted and she found she had missed being a part of their group. But she felt uncomfortable and false, so she deliberately tried sex, only to find that she didn't like it. Her fear of having gay tendencies is becoming an obsession, and, added to that, she realizes she's pregnant. She doesn't want to get married and abortion would be no simple matter in her family. Her life is a mess and there is no one she can talk to. Is suicide the only answer?

Grandmother: Sixty-five-year-old woman who lives near her children and grandchildren and loves them deeply. But they are busy living active lives and she has been lonely. Now she is thinking of doing something that's guaranteed to alienate them. She has formed a warm friendship with a gentleman friend and they are thinking of living together without benefit of marriage. If she got married she would lose her first husband's Social Security check, and she and her friend would not have enough to live on without it. Would she miss her relationship with her family and be lonelier without them than she is now? Pricilla is her favorite and not as conservative as the rest. But will her tolerance of the new ways extend to her grandmother?

[As the scene opens, Pricilla has just come home from school to find her grandmother wandering around the house wondering where everybody is.]

Pricilla: Oh Gran, I'm so glad you're here! [hugs her with great enthusiasm] The family has gone to a potluck supper at the church. We'll fix something for ourselves and have a ball!

*Role-descriptions and dialogue are for the conductor only; they are given as guides.

33

Grandmother: Bless your heart, my Dear. [hugs her in return] It's so good to be here with you!

Pricilla: C'mon into the kitchen. We'll start things rolling. [leads the way to the kitchen]

Grandmother: [following her into the kitchen] How come you aren't with the family tonight?

Pricilla: I knew I'd be late getting home. [pause, low tone] That was just an excuse. I didn't want to go.

Grandmother: There must be a good reason why you didn't want to go. Do you think you could tell me?

Pricilla: [getting out some pans] I've got some problems that don't seem to have a solution.

Grandmother: In the past we've sometimes been able to figure out a way together. Remember that Girl Scout camp you wanted to go to? [pause, gentle tone] Do you think you might try telling me?

Pricilla: [moving pan to stove and emptying the contents of a plastic dish into pan] I don't know, Gran. I think this is somehow out of your league.

Grandmother: It could be a problem I haven't experienced. [pause] Even when I haven't known the answer, I've known where to find it. Haven't I?

Pricilla: Yes you have. [long pause]

Grandmother: Once when I was a girl and out with a new boy, I was with a group I didn't know and frankly didn't like. The petting went on quite far before I could stop it. A couple of weeks later my period didn't come. I was frantic. I wasn't sure anything had happened, but then I wasn't sure it hadn't!

Pricilla: [with obvious interest] What did you do, Gran?

Grandmother: I went to the library. I wasn't sure the librarian believed my story about a term paper but I didn't care. After a long search I found my answer. It was probably anxiety that caused the period to be late.

Pricilla: Gosh, Gran, do you think the library would have anything about homosexuals?

Suggestions for the Roleplay "This Crazy World"

Selecting Your Players. Youths of any fighting ability can play the roles. The part of the grandmother is not a prop role and will bring out the Parent Ego State of the player. This is a role without conflict and the emphasis is on sympathetic listening and understanding.

34

Instructions for Players. Personalize your remarks to your players and let them feel your sympathy. To the grandmother you could say: "This younger generation is supposed to be open-minded about sex life, but does that extend to having a grandmother living with a man without sanction of marriage?" Etc. Inform the grandmother-player that she has a good relationship with Pricilla because she has used her own experiences from her early life to help the girl. But caution your player against giving judgmental examples. Finish your instructions with a remark that brings your player right into the act like: "You have gone over to your daughter's house to tell the family of your circumstances, only to find no one home. You are about to leave when Pricilla walks in."

To help Pricilla you could say: "You've tried but found you couldn't talk to your father or mother. And your minister? He's living in the dark ages and probably doesn't know what the word *gay* means." Etc. Then set the scene: "You have come home from school to find your grandmother wondering where everybody is. The family won't be home for some time—they are at a potluck supper at church—and you are glad. Your grandmother is a great gal and you have a good relationship with her."

What to Watch for. Leveling is important here and it may be difficult for the Pricilla-player to bring out all the facts. Watch for facts that are ignored or brought out reluctantly. The emphasis in this role is on sensitive listening and gut-level sympathy. Is the grandmother-player helping the girl-player to feel enough at ease so that she can level? Is she giving some experiences from her own life or that of her friends? Do they apply to the age of the granddaughter?

The start of the role-conversation here shows Pricilla's love for her grandmother, and it sets the tone for a good confidential talk. This is one way an outgoing personality can help a type who finds it hard to show feelings. The leveling in the role-dialogue given shows how the problem can be brought into the open quickly. Sensitive listening, gentle persuasion, and physical contact are also shown in the role-dialogue here.

Discussion Following the Roleplay. Find out from the girl-player whether or not she was feeling the sympathetic listening of the grandmother-player. Did the grandmother-player use any physical contact? If not, why not? Did the grandmother-player's examples of experiences from her early life (or experiences made up from a friend's experiences or knowledge gained in reading) help the girl-player? Were all the facts brought out? If some facts have been forgotten, is there a reason? If it can be done in a helpful way discuss abortion and suicide. Discuss homosexuality. The role tries to show that a lack of knowledge about what is "gay" is one reason for the problem. Games at the end of this chapter and chapter 6 on love can be used to advantage here.

MIMIC CHARADES

Game 1

"Body language," that is, a player's expression and his body movements, can tell a great deal more than people realize. Your eyes may be getting a more accurate message than your ears. Sensitive listening includes receiving this non-verbal communication. Here is a game to help players become sensitized to this form of listening and to be aware of all the ways in which we express ourselves. It is a variation of charades.

The game is played by choosing up teams of from five to ten players. (If the group is very large the teams may be larger.) The teams should have a name and a motto to intensify the competition.

When teams are ready each team sends a representative to the conductor. He then gives each representative a word which they take back to their teams and act out in pantomime. Unlike regular charades, the players will be unable to use the sign language for syllables (long and short, cut-off, and so on). The word must be shown entirely by actions or facial expressions. When a team has guessed the word, the representative quickly raises her hand. The first team to guess gets three points, the second team two, and the third team one.

Then another representative from each team goes up to receive a word, and so on. When one team completes a set of words or reaches twenty points (or any other amount the conductor chooses), the game will be declared at an end and the winning team given proper commendation.

Example: After receiving the word, the representative will go back to his or her team and demonstrate picking up a jar and screwing off the cover. The player sets the cover down and reaches into the jar, bringing out something she puts in her mouth. Her face will register a sour taste—but one she likes! The word she had been given is, of course, *pickle*. (This is only one of many ways the word can be acted out.)

The conductor can select any words and any number of words for the game. They can relate to each other or (for more difficulty) be unrelated. The conductor should start with words easy to mimic and then increase their difficulty.

Some sets of words are (pick some from each list for an unrelated set):

violin	basketball	pickle
accordion	tennis racket	lemon
piano	skates	bitter pill
drums	baseball bat	spoon of castor oil
tambourine	golf club	onion
bass fiddle	sled	corn
xylophone	baseball	apple
organ	hockey stick	peach
saxophone	golf ball	walnut
banjo	skis	a grape
zither	ski pole	mango
boots	rocker	clown
pantyhose	typewriter	acrobat
girdle	car	lion tamer
belt	chair	ringmaster
coat	door	circus horseback rider
overshoes	book	circus barker
bathrobe	bed	horse rider on merry-go-round
parka	lamp	tightrope walker
wet socks	clock	Ferris wheel rider
purse	dictionary	trapeze artist
dress	table	circus

I-THOU FEEDBACK
(An Active Listening Game)

Game 2

AIM: To promote better listening.

PLAYERS: Pair players off, preferably with a boy and girl in each pair. Place your players in chairs facing each other.

SETTING: Pleasant room devoid of distractions such as disturbing noises, drafts, room too hot or too cold, etc. (Or as close to this as possible.)

Instructions for Speaker:

Have the speaker choose a subject or topic that he feels deeply about, and which has some emotional overtones for him (if possible). Or have the speaker choose a controversial subject on which he has a strong viewpoint.

Examples: Living together without marriage
Cheating on tests
Abortion for teenage girls
Use of the pill for teenagers
Dirty politics
Double standard for premarital sex
Legalization of marihuana
Older brothers and sisters

Instructions for Listener:

This is a game on listening so impress upon your audience the need for listening carefully to all your instructions. These are important because they are blocks to listening.

1. Explain that most physical distractions have been eliminated, but that players will have to use rigid discipline for a short time to rid themselves of other distractions, such as:

    a. Watching clock, fiddling with keys or other objects, pencil tapping, etc.
    b. Desire to daydream
    c. Desire or compulsion to think of problems of the day or worries, etc.

The listener should:

2. Give full physical attention. He should sit up straight and look directly at partner. This is supposed to keep a listener alert and free from distractions.

3. Watch for the "body language" of the speaker, his gestures, grimaces, and the pitch, tone and volume of his voice. Words alone do not tell the whole tale.

4. Note any omissions the speaker makes since understanding often comes from these omissions.

5. Be as open-eared and as open-minded as possible. The listener must rid himself of any negative feelings he might have for the speaker.

6. Gain as close a rapport with his partner as he can. He must try to achieve empathy! Have him make a heroic effort to understand what the speaker is saying (or trying to say!), and why he feels the way he does.

7. Be careful to avoid being so overstimulated by a word or so repelled by a point of view that he forgets to listen.

8. Be sensing and interpreting as the speaker-partner is talking, so that he can demonstrate acceptance in the form of feedback. This does not mean he has to agree with him, only that he show his partner he is listening and is "with him" and "on target."

9. Keep in mind the purpose for listening. In this case it is the gist of what the partner is saying (or trying to say), and the reason for the emotion behind it (if there is any and if it can be figured out).

First step is the SPEAKER'S DISCOURSE:

The conductor will explain that the speaker will have about a minute to talk on his favorite subject, during which the listener will give his full attention. After the minute is up the conductor will call, "time's up!"

Second step is the FEEDBACK that the LISTENER will give:

The listener will give the speaker feedback on what he thought the speaker was saying and meaning. If emotion is observed, and the listener has some understanding of what caused it, he should try to indicate that he understands the reason for it. Encourage the listener to achieve empathy if possible.

Thomas Gordon, who developed "active listening," advises his participants to avoid these typical twelve points in using feedback technique:[2]

1. Ordering, directing, commanding.
2. Warning, admonishing, threatening.
3. Exhorting, moralizing, preaching.
4. Advising, giving solutions or suggestions.
5. Lecturing, teaching, giving logical arguments.
6. Judging, criticizing, disagreeing, blaming.
7. Praising, agreeing.
8. Name calling, ridiculing, shaming.
9. Interpreting, analyzing, diagnosing.
10. Reassuring, sympathizing, consoling.
    ("Things will be better" or "I used to go through that too.")
11. Probing, questioning, interrogating.

12. Withdrawing, distracting, humoring, diverting.
("Just forget about it" or "Let's talk about something else.")

Third step is the SPEAKER'S REACTION:

When the listener has given his feedback, the speaker will correct it or indicate it is all right. He will then rate the listener with one or two points for good and very good feedback, or four points for excellence.

Then the players will reverse roles and repeat. After each player has had a chance to be both speaker and listener, the players will shift chairs to get new partners, and then repeat the game. This can be repeated as long as the game holds interest and time permits. If a player complains that a partner is not being fair with his points, a game can be played with a third person as an observer.

The player with the most points wins.

Listing as a Discussion Technique:

LISTENING
(to be used for the roleplaying discussions)

A. When does a person listen best?
- *1. When he is interested in the subject or has a compelling reason to know information.
- 2. When distractions are minimized or when he has trained himself to eliminate distractions.
- 3. When taking notes. Why?
  - a. Because the act improves attentiveness.
  - b. Listener is able to review what he has heard.
  - c. Notes remedy weaknesses in listener's ability to learn from spoken word.

B. List bad habits that discourage listening.[3]
- 1. Faking attention and not trying to understand.
- 2. Being so concerned with facts that you miss what they're leading to.
- 3. Trying so hard to memorize some facts that you miss others.
- 4. Avoiding difficult listening. (TV-type of listening is so habitual that you're unwilling to make extra effort when more is required.)
- 5. Dismissing the subject as uninteresting.
- 6. Using speaker's appearance or poor delivery as an excuse for not listening.
- 7. Yielding to distractions. (Draft, noise, etc.)

C. Why is *critical* listening difficult? (Why is it hard to combat salesmen, rumors, politicians?)[4]
- 1. Time element. No time to analyze.
- 2. Oral persuasion more effective.
- 3. Spoken word not as accurate. It can be colored with humor or pathos.
- 4. Drop defenses in a chance encounter; there is no motive to be critical.
- 5. Person-to-person relationships promote trust.
- 6. Blinded by labels that produce emotion.
- 7. Intrigued by idea of exclusive information or "inside dope."
- 8. When hearing many good points, you forget to be critical or look for bad points.

*The answers given are only suggestions. Have your players give their own answers, and do not give suggestions unless it is absolutely necessary.

9. Tendency to follow the crowd.
10. Sympathetic to someone like us. A "good Joe."
11. Attracted by the glamor of someone well known in a field or an authority on a subject.
D. List barriers to listening.
1. Words have different meanings to different people.
2. Language and other differences in social or cultural backgrounds.
3. Not trying to understand. Tuned out.
4. Have values different from speaker's. (Object to the views expressed.)
5. Dislike speaker or feel resentment toward him.
6. Words loaded with emotion tend to tune you out from rest of speech.
7. Speaker has nervous mannerisms. (Jingles money, glances away from audience, etc.)
8. Speaker doesn't speak clearly. (Poor enunciation, mumbling, slurring, etc.)
9. Speaker talks too softly or too loudly, or in an annoying tone.
10. Listener has preconceptions, lacks attention, notices interruptions.
11. Objects in speaker's mouth make it hard to understand. (Pipe, cigarette, gum, etc.)
12. Speaker doesn't look directly at listener or audience.
13. Speaker's words are not meaningful to listener.
14. Speaker sends indirect messages that repel.
15. Speaker uses put-downs, judgmental statements, sarcasm, name calling, prejudices that produce anger.
16. Speaker shows dogmatism or argumentativeness.
17. Speaker talks too much and too fast.
E. What are the three most important things for listening?
1. *Discipline.* Listening is an art like music or painting and the discipline must come from within. It is necessary to listening.
2. *Concentration.* Listening is hard and demands an active participation and a relaxed alertness.
3. *Comprehension.* Necessary to understand and grasp the idea or meaning of what is heard. This should be the intent of both the sender and receiver, for a good listener joins the speaker in the excursion of understanding.

Take Note:

These listings are to help the teenagers form an idea and express it as concisely as possible. Too often teenagers' stumbling attempts and vague ideas are met with put-downs or ignored. The conductor can pick up these vague ideas, and (with or without the help of the suggestions) bring the ideas into focus with the question. He can then write them briefly on the blackboard.

Haim Ginott, in his book *Between Parent and Teenager*, says we: *win a teenager's attention* when we listen and respond sympathetically, *win his heart* when we express for him clearly what he said vaguely, *win his respect* when we are authentic and our words fit our feeling.[5]

Suggested reading:

Thomas Gordon, *P. E. T., Parent Effectiveness Training* (New York: Peter H. Wyden, Inc., 1970);
Ralph G. Nichols and Leonard A. Stevens, *Are You Listening?* (New York: McGraw-Hill Book Co., 1957).

4

understanding

Failure to communicate is often—too often—caused by a lack of understanding. Paul Tournier, in his book *To Understand Each Other*, tells us that to understand is to love and that a person who feels understood feels loved.[1]

The roleplays in this chapter attempt to get at some basics of understanding by showing how everyone approaches problems from a different angle. Remember the story of the four blind men approaching an elephant from different angles. Using their hands to see instead of their eyes, one described his trunk, another his ears, and the other two his body and a leg. Although their descriptions varied a great deal, together they gave a fairly accurate picture of an elephant.

The roles in this chapter aim at getting the players to look at problems from perspectives other than their own. To do this the player is put in a situation where the conflict is best resolved by understanding something of his opponent's way of looking at an issue.

EXTROVERT *vs* INTROVERT

Explanation of Types:

EXTROVERT:[2] Extroverts like people and like to be out in the bright lights with crowds, gaiety, and movement. They always have something to say and love to retell a story. Absorbed in events and in getting things done, they sometimes overdo almost to the point of exhaustion. Extroverted types work out their frustrations by action. They are sociable, cheerful, friendly, and tend to take a positive attitude. An extreme extrovert tends to avoid introspection and may even think it unhealthy.

INTROVERT: Introverts keep their best qualities to themselves and only show their gifts in sympathetic surroundings. They are over-conscientious and tend to be pessimistic, reacting to a new idea with a definite "no."[3] Introverted types are happy sitting alone in the dim light of an open fire or curled up with a good book. They look inward at their own ideals and goals. Interested in what an event really means, they love to evaluate it. They are the inventors, planners, and theorists. An extreme introvert is a passive person and tends to put off or avoid action.[4]

Roleplay 5 (the above Types personified in a situation):

*Elmo: A high school senior running for president of his class. He has enjoyed the campaigning, going around greeting old friends and making new ones. His girl, Lee, whom he had taken with him wherever he could, has been acting strange. He missed her the other night and found her out in the garden—alone! And the weather was cold! She sure was a strange girl, but she fascinated him.

Lee: A sophomore who has been very proud to be known as Elmo's girl. But she can't stand any more parties! She has a desperate need to relax and collect her thoughts. She enjoys being with Elmo, but whether he joins her or not, tonight she is determined to stay home.

[Elmo walks into Lee's house without knocking (as he always does) and late (as he usually is). He is horrified to find Lee in a housecoat.]

---

*Role-descriptions and dialogue are for the conductor only; they are given as guides.

Elmo: Gosh, Lee, don't you know we haven't got a second to spare? You're not dressed!

Lee: I'm not going, Elmo. With school every day and a party every night, I haven't had a chance to catch my breath, much less think. Oh Elmo, it'll be fun to stay home tonight. We haven't had a chance to talk together, alone, for a long time. We could have a fire.

Elmo: (starts to move around and talk rapidly) So we've been out a lot lately. You sound as if parties were hard work!

Suggestions for Roleplay "Extrovert vs Introvert"

Selecting Your Players. Selection is important in this role, because one who understands how an extrovert or introvert feels will more readily fight for that viewpoint. The selecting can be done in different ways. The simplest way would be to read the explanation given for the extrovert and introvert (but not the role-description or role-dialogue!) and have members of the audience volunteer for the part that seems closest to them. (You should explain that most people have some characteristics of both the extrovert and the introvert but more of a tendency towards one or the other.) For this role the players should be of equal fighting ability if possible, but identification with the role is more important.

Instructions for Players. You will be addressing each player in your own words and in the second person, so with "body language" and tone of voice you should instill some sympathy for the individual role he or she is to play. Part of what you could say to Lee, the girl part, might be: "Does he really love you if he would rather spend his time at parties talking to other people?" And to bring her into the role say something like: "You hear the car, the door slam, and here he is in a hurry as usual." To Elmo, the boy part, you could say: "From something Lee said, you gather she isn't too crazy about going to the party tonight. What cooks with this woman? She objects to parties and having fun?" To bring him into the role you could say: "You walk into her house and there she stands obviously not ready for the party."

You are trying to heighten the conflict between the players. Think of the program on television some time ago where the moderator would give his contestants conflicting stories. Before he threw them into the situation, he would turn to the audience and say, "Aren't we devils?" He knew what would bring controversy. If you know your players, there is a great deal you can add to increase the conflict.

What to Watch for. In this role each player should be working toward understanding the other person. Each should be trying to see how the situation looks from his opponent's point of view. They should work toward some kind of compromise where both can be happy. Take note of manipulating tactics. Look for both the hard and soft sell. Watch for leveling. When both can state how they feel, the role will move faster. Your players may not understand each other and may not want to compromise. This need not continue too long, for you can stop the roleplaying and ask the players how they are feeling. If their trouble lies in not seeing each other's viewpoint, you might model for the player needing help. You could show how a great deal more can be accomplished by using understanding and some willingness to compromise. Nevertheless, the person may still not want to compromise. Don't push it too hard; nothing should be forced in roleplaying.

Discussion Following the Roleplay. After the roleplay, you should ask each player in turn how he or she was feeling during the playing and at the end. Encourage your players to bring up any problems they have encountered. Occasionally a player suffers a crushing defeat through no fault of his own but simply because the other player is rigid and will not budge. If he feels sensitive about it, he should be reassured, for he has been under a handicap. Remember that in each playing of these roles you are encountering something different. This is because the players are different. Their reactions to what is said (triggered by their past experience) will be unpredictable.

If you have two casts playing the role, you can ask the audience, after the second playing, to compare how the different players handled the situation. Did the players level about how they really felt? Did each one try to understand his opponent's view or did they all only cherish their own? Did any of them impose his views on the other player?

Tournier, in his book *To Understand Each Other,* says that men are attracted to their opposite in a woman (and *vice versa*), and that one of the purposes of marriage is to allow the two, through their close relationship, to discover and understand what they have not known before.[5] The audience may wish to test this with their own experiences of what attracts them to the opposite sex.

48

INTUITION *vs* SENSATION[6]

Explanation of the Functions:

INTUITION: Intuition tells you "what may be" and Uncle Hunch will play this part. He is an adventurous fellow alive with new ideas that have unlimited possibilities in the near future. At home with the invisible perception of ideas, he just seems to know something is right and reasons from there. Security (in the sense of certainty of outcome and realized potentials) stifles him. He has instinctive or psychic insight which he uses for his inventions.

SENSATION: Sensation tells you something "exists." Aunt Fact will take that role. She is an accurate observer who is convinced of the reality of the five senses. Being a worrier, she must be absolutely sure of any new undertaking or she is frantic with anxiety. She will scientifically bring in all the facts before any decision is made. She is a practical person.

Roleplay 6 (the above functions personified in a situation):

*Uncle Hunch: He has decided to sell some income property because he is off to new horizons and property is just a drag to take care of. It interferes with his work on his inventions.

Aunt Fact: Wife of Uncle Hunch whose name is Constance. She knows that this income property takes very little time to take care of and furnishes security for the present and future. She is in a very real panic at the thought of its being sold. He needs her signature on the deed to the property.

[Hunch has just come in the kitchen door all excited because he has a buyer for his income property. Aunt Fact continues with the dishes.]

Uncle Hunch: Constance Honey, I've got a buyer for the apartments!

Aunt Fact: Oh Hunch, don't you think we ought to talk this over a little more? I've been reading some statistics in the paper . . . .

Uncle Hunch: I don't care about any old statistics. I have a whole new project coming up and I can't be bothered playing nursemaid to apartments.

Aunt Fact: You seem to feel very strongly about this.

Uncle Hunch: I do, Constance. Those apartments stifle me. I need to be free.

---

*Role-descriptions and dialogue are for conductor only; they serve as a guide.

Aunt Fact: I çan see that. What if I were to help with them? Doing the cleaning and renting?

Uncle Hunch: Well . . . . Why should you want to?

Aunt Fact: I have a desperate need for security, Hunch, and those apartments furnish it.

Uncle Hunch: That's foolish, Constance. You know that!

Aunt Fact: Maybe. But that's the way I feel. I can do the painting, Hunch. We could hire someone to do the things I couldn't.

Uncle Hunch: There's no need for you to work. I'll see that we have enough money! This new invention—

Aunt Fact: Where have I heard that before!

Uncle Hunch: So you've heard it before. My mind's made up. I'm going to sell whether you like it or not.

Aunt Fact: Have you forgotten that you need my signature on the deed?

Suggestions for Personification Roleplay "Intuition *vs* Sensation"

Selecting Your Players. The selection of players is also very important here because the player who has a natural feeling for his role can more easily fight for this view in a conflict. If your players are having trouble understanding what the Intuitive Function is, even after you've read the explanation of the Function, you might suggest some TV detective heroes who go about solving mysteries by playing a hunch. Dr. Gannon on *Medical Center* often uses educated intuitiveness in diagnosing his patients. And, just as often, he is in conflict with a doctor who is more concerned with just the facts before him. This latter doctor, who feels that Gannon's diagnosis is not warranted by the facts, is using the Sensation Function. The player's identification with the role is more important than his fighting ability.

Instructions for Players. In these personifying roles, you should try to make sure the player feels that the Type or Function he is playing is worthwhile and something he is proud of. In this particular role the Intuitive Function is being shown in the person of an inventor; he is unconcerned with security and hardly knows the meaning of the word. He represents the extreme. Our society has a strong tendency to belittle or dismiss impractical inventors, so your player may feel uncomfortable taking this part. (On the other hand, it has become fashionable to scorn middle-class securities, and adolescence is naturally a rebellious age, so your player may feel comfortable in the role.) If you sense (here, or with any of the personifications) that your player is feeling anxious or insecure because of the role, take time to dispel such a feeling. Explain that the capacity to go out to a problem in an intuitive way is useful in all creative work, especially quantum physics. All innovators and inventors should be considered in the same light. You could also bring up the detective heroes, Dr. Gannon, etc.

To help Hunch, you could say: "What's with this woman? Does she want to tie you down to some stupid apartments when you have this important invention that needs your whole effort? Doesn't she understand your need to be free?" Remember to bring him into the roleplay with something like: "You have just come from the realtor's and you've found out they may have a buyer for the apartments. You open the door and hurry into the kitchen where you know you'll find Constance. There she is and you can't wait to tell her the good news."

To Aunt Fact, who is a practical person, you could say: "Doesn't Hunch understand your need for security? He's always off with something new like a little boy with a new toy." To bring Aunt Fact into the role: "You are doing dishes in the kitchen worrying about the apartment building's being up for sale. You hear Hunch's car and you can tell by his step he's happy about something. He bursts into the room, all smiles."

Bear in mind you are only trying to provide as much conflict as you can. It is up to the players to deal with the conflict and come up with ideas for compromise.

What to Watch for. In this role each player should be working toward understanding the point of view of the other person. This does not mean giving in to the other player's demands without fighting for some rights of your own. Hunch should understand his wife's need for security and try out some things in the way of compromise. Aunt Fact should realize Hunch's need to be free to go on to more creative ideas, and she should offer to compromise if he will forgo selling. Each should take a try at understanding and be able to give a little. Is either player taking an uncompromising attitude that cuts off dialogue? In the role-dialogue given here Hunch takes a hard line—almost with the first statement. Aunt Fact counters this beginning with calm, and with active listening, giving Hunch the freedom to level about how he feels. Watch to see if each of your players levels about how he or she feels so that the other player can begin to understand.

Is either of the players so interested in his own view that he doesn't listen to his opponent and sense his opponent's needs? When Hunch, in the role-dialogue here, brushes aside Aunt Fact's leveling, he is being insensitive to her needs. Notice that when Aunt Fact offers to compromise it takes some of the steam out of Hunch's aggressiveness. Watch for times when one player is trying hard to think of compromises that will bring about a solution and his opponent is not trying at all.

If both players are stubborn and the roleplay gets bogged down, stop the playing and ask the audience if they can see what has happened. Ask them if they can suggest what might be done. If no answer is forthcoming, you might have to suggest that there is some way both could be happy. You may have to model to show them.

Discussion Following the Roleplay. Discussion is always necessary after role-playing. It becomes more important if the players don't try to compromise. Players and audience should be helped to realize that true dialogue is a free exchange of ideas and that, if dialogue is to go on, participants must accept new viewpoints. Barriers must somehow be surmounted. The players should understand that they are looking at a situation from different angles, and each should see that both his viewpoint and that of his opponent may be only partially right.

Discuss "intuitiveness." This is a gut-level feeling of knowing something but not how. If there were instances of good leveling and non-leveling, bring these up so that the players and the audience can realize the difference it makes. Was there any outstanding manipulation? Were there put-downs when a player was being aggressive or defensive? The role-dialogue here shows Aunt Fact using sarcasm for a put-down when she is on the defensive and it gets a negative result. If there have been any instances of negative results from put-downs by either player, discuss them.

THINKING vs FEELING[7]

Explanation of Functions:

THINKING: The Thinking Function tells you "what" a thing is. Miss Hair will take this role. She attacks all problems logically, identifying everything in a precise way. Because she is afraid she cannot control her emotions, she severely represses them. She analyzes facts and has the need to mold or synthesize them into some kind of a pattern or idea. She is the reasoner; logic is her god.

FEELING: Feeling tells you what a thing is "worth" to you. That part is played by Mr. Cool Cat. His feelings are rational and he has them under control at all times. Understanding people and what they want, he uses this tactfully to gain what he feels is right. He is a people's person. Judgment of value is his "thing."

Roleplay 7 (the above Functions personified in a situation):

*Mr. Cool Cat: Hippie sort of character who teaches art at the Hilldale High School. He has used his keen knowledge of humanity to sway many of his co-workers into a teachers' strike for higher pay. He feels values in our society have gone out of kilter. He welcomes the chance to bring to the notice of the community the fact that plumbers are paid more than the teachers of their children.

Miss Hair: Miss Hair is the high school math teacher in this small town. She feels strikes hurt not only the children but also the capacity of the teachers to control and teach. How can you teach your pupils to talk out their problems in an orderly way when you yourself are striking rather than trying to negotiate?

(No dialogue is given with the roleplay; for a guideline, the conductor can use the dialogue to the Open-End Play, which is found after the roleplay suggestions.)

---

*Role-descriptions are for the conductor only.

Suggestions for the Personification Roleplay "Thinking *vs* Feeling"

Selecting Your Players. How you select players will determine the ease with which this role is played, for it is one of the more difficult roles. If you are acquainted with your audience, you may wish to select the people you feel can best identify with the parts. If you don't know your audience well, read the explanations of the Functions (thinking and feeling) and ask for volunteers who can identify with each part. You should explain that the roles given here show the extreme of the Functions. The ability of your players should be fairly even although in interpersonal relations a person who closely fits the "Feeling" description usually has a slight edge. If you think one of your players is over-matched, pep-talk him or her into greater fighting power.

Instructions for Players. You will be trying to heighten the conflict between the players, so sympathize as much as you can with the player you are addressing and the role he is to play. To Mr. Cool Cat you could say: "You realize Watergates and useless wars will continue until people get their sense of values into clearer focus. What's the matter with some teachers—can't they see that the issue here is more than salaries? Like Miss Hair! She's too rigid. Maybe you can help her open up and live a little!" Etc., etc. To Miss Hair you could say: "Don't these striking teachers realize what they're doing to the children? With the emotions of the whole town aroused, violence could break out at any time! Especially if new teachers are brought in as strike-breakers." Etc.

Don't forget to bring your players into the scene with something immediate. To the Cool Cat player you could say: "Well, well, here comes Miss Hair, the oh-so-proper math teacher. You wonder if she's going to join the strike. That would be news! You go over to greet her." Etc. To the Miss Hair player you could say: "There's Mr. Cool Cat. He's magnetic, all right—personable, attractive, really quite likable. But he's also the leader of this terrible strike. Maybe you can help him see the dangers he's leading the teachers into. It looks like he's coming over to greet you." Etc.

What to Watch for. The aim of the roleplay is understanding the opponent's viewpoint. Watch how the players approach this. The dialogue of the Open-End Play (provided to help the conductor but not the roleplayer) shows something of how these two types of people think and react. The roleplayers may or may not bring this out, depending on how much they use the Function. Watch to see if they stick to their roles and bring this out during discussion. Watch for leveling. If there is leveling the roleplay will move faster and each player will have an easier time understanding his opponent's view. Take notice of when a player is not leveling. Notice when a player can concede a point (as Cool Cat does in the dialogue) and commend the player during the discussion period. Take note of the player who is willing to apologize when wrong (as Miss Hair does about the reason teachers stay in this town). Watch for put-downs that slow the roleplay.

Notice when different types of manipulation are taking place. Both the manipulator and the manipulated may be unaware of this.

Discussion Following the Roleplay. Discussion is especially important with these personifying roles, because the players must be made aware that each of them is simply using a different approach. They should realize that each approach is a valid way of looking at a problem. If both players can understand that each is simply using his or her own best way, they may not spend so much time trying to change their opponent—not only in the roleplay but in their everyday personal relations. If a player has not been able to stick with his role, he may be using his opponent's approach (or another one) in his everyday life rather than the one for the role he has been trying. This is not always easy to see until it is tried.

The game "Who Is Me?" (p. 58) makes use of the Functions (Sensation, Intuition, Thinking, Feeling) and Types (Introvert, Extrovert) described in this chapter and can be played either before or after the roleplays. The game can be used to give everyone—both roleplayers and audience—a chance to determine which Function and Type he or she feels most at ease with.

If there has been manipulation among your players, you might like to have them look at another way of reacting (see "Contrasting as a Discussion Technique," p. 70). If there is enough interest, your group may want to play the manipulation game on page 63.

THINKING *vs* FEELING

Open-End Play:

(Open-End Plays are not roleplays but simply short skits that make no attempt at an ending. After play-reading this skit you can discuss possible endings, but that isn't necessary. The objective of an Open-End Play is to provide a medium for discussion. In this skit the interaction between the two Functions, Thinking and Feeling, is shown to bring out points necessary for understanding and discussion. If for some reason both the roleplay and the Open-End Play are needed, use the roleplay first so as not to inhibit the creativity of the roleplayers.)

[Scene: Strikers are demonstrating in front of the Hilldale High School in a small town in Wisconsin. Mr. Cool Cat, who is among them, sees Miss Hair walking toward the group. Knowing her feelings about strikes, he wonders what she is doing here. He goes forward to greet her.]

Cool Cat: Well, Miss Hair, have you come to join our little strike?

Miss Hair: Surely you know better than that! I'm hoping to convince some of these teachers of the falsity of their thinking.

Cool Cat: Are you sure it's thinking that persuades them to be a part of this agitation? Couldn't it be an instinctive feeling for what is right and worthwhile? Why bring thinking into the picture?

Miss Hair: My dear Mr. Cool Cat, thinking is responsible for the technological miracles of our civilization. Why go back to the cave age?

Cool Cat: You mean it's better to kill a man with an atomic bomb or flamethrower than to just hit him over the head?

Miss Hair: No, of course not. But what has that got to do with this strike?

Cool Cat: Our civilization amounts to nothing without a clearer picture of what is worthwhile and what isn't. Here in our own bailiwick, it's higher wages for teachers. We somehow must get the school board to listen to our pleas.

Miss Hair: But didn't most of the teachers come to this small town because they were inexperienced and probably couldn't command the salary of a larger school?

Cool Cat: Speak for yourself, Miss Hair. Many of the teachers, including me, are here because this is our home.

Miss Hair: I'm sorry—I didn't know that. But if salary is so important, couldn't you go where you would get higher wages?

Cool Cat: Yes, we could. But we like this town. To me, teaching is more important than plumbing. But here plumbers are paid considerably more than teachers are. Low salaries do not encourage the best caliber of teachers.

Miss Hair: But is disruption the best way to help? Have you thought this over carefully enough?

Cool Cat: You bet we have! Everyone involved in this strike risks not only his job here but rehiring in the profession.

Miss Hair: I know! So why risk it?

Cool Cat: This technological age you talk about has advanced without a decent sense of values. There isn't enough emphasis on what is really worthwhile. Someone has got to take risks to make this town aware of what it's doing.

Miss Hair: But the risk you're taking is too great! And is it really good for the children?

Cool Cat: I sincerely believe it is. Our children should be made aware of the lopsided sense of values. They should be encouraged to take risks for what they can see is right.

Miss Hair: But that's just the point! Will the children see the issue or only notice your methods?

Cool Cat: What do you mean?

Miss Hair: What if some of these children become plumbers and remember the strike only as a means of getting higher pay? Won't that undo what you hope to achieve?

Cool Cat: You have a point there. They should be made aware of the real issue!

Miss Hair: Wouldn't it be better just to arbitrate?

Cool Cat: Arbitration is failing, Miss Hair. We need action!

Miss Hair: But strikes arouse negative emotions. Already people are talking about replacing the striking teachers. There could even be violence!

WHO IS ME?

Game 3

AIM OF THE GAME: To help players get acquainted.

To help players become aware that their friends and associates may approach problems from a different angle. Implies people are unique.

PLAYERS: If the group is large it would be best to break into groups of from 4 to 6 players. A group leader, who has had some briefing, should then be provided for each group.

MATERIALS: Blackboard or easel. Photocopies of the explanations of Types and Functions found on pp. 46, 49, and 53. Photocopies of the descriptions of Superior Functions in text of game, and photocopies of the Function Wheel.

This is a game for Carl Jung buffs. Carl Jung believed that because of unique conditions, from childhood on we tend to become more and more comfortable and more and more skilled in approaching a problem in a certain way.[8] This is done subconsciously and often we are unaware of how we approach things.

Young people are often upset when an admired parent, an older brother or sister, or even peers ask questions like: "Why have you always got your head stuck in a book?" "Why are you always going off half-cocked? Can't you lay out the facts first?" Young people are not always aware of why they approach a problem the way they do. They do know that when they try to do the things the way their parents or peers advise, they sometimes become confused. Young people must be helped to realize that perhaps their own way *is* best. They may be unconsciously following the pathway of a Function or Type and no one way is any better than the other. Each is a valid way of adjusting to life. Moreover, the way they have unconsciously picked has proved the best way for them. It is wise to avoid an extreme in any direction, but each person is unique and must be allowed to be himself or herself. They should be proud of saying, "I gotta be me!" But who is me?

Part 1. It would be best to start with the Types, *Extrovert* and *Introvert*. The group leader should have an assistant read the Type descriptions from the photocopied explanation of Types while he jots down the points for all to see on a blackboard. He will explain to the group that all will have some characteristics of both but a tendency toward one more than the other. If this is to be used as a get-acquainted game, the leader will give his name, and some facts about himself, then mention the Type he thinks he might be.

If he chooses Introvert, he might like to tell this story: (Extrovert could tell the story from his own point of view) My older brother, who is now a successful salesman, took an occupational test that asked the question: If you had your choice of reading a good book or going to a party which would you choose? My brother thought it a silly question—naturally everyone would want to go to the party! My brother was an Extrovert, always asking me, "Why do you always have your head stuck in a book?" Well, I now have confidence enough in myself to answer, "For the same reason you always like to be with a group of people. We're different!"

The group leader should then allow each one in his group to follow his example of giving his name, some facts about himself, and his Type. They may want to stop here, or go on with the Functions.

Part 2. A. The Functions are more difficult but well worth the effort. The leader should start by drawing a circle and putting in the four Functions, pointing out their relative positions. He will then draw another circle and put in just one Function. Suppose the group leader uses *Thinking* as his Superior Function; it's the Function he uses the most and so easily that he's almost unaware of using it. On the upper quarter of the circle the leader will write "Thinking." As his assistant reads the explanation of the Function from the photocopy, he will jot down descriptive words. He could explain his own experiences with this particular Function or use the example (or have his assistant read the example) provided below:

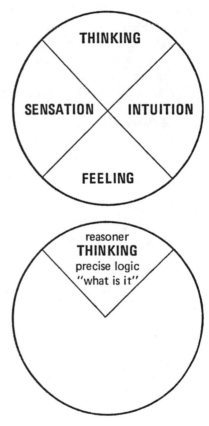

"I attack a problem by first figuring out just precisely what the problem is and then reason it out in a logical way. I sort out available data; then I analyze and synthesize it."

59

B. The group leader should then take the opposite Function on the wheel, which in this case is *Feeling*, drawing the circle and putting in the descriptive words as they are given. He should explain that for HIM as a Thinking Type this is his Inferior Function because he uses it so seldom. After all you can't be feeling something and reasoning it out at the same time! Knowing that he represses the Feeling Function has helped him to realize that Thinking, the opposite on the Function Wheel, was his Superior Function. Although he represses the Function, he knows his sister uses the Feeling Function as her Superior Function.

"This sister of mine seems to know me better than I know myself, and everyone else for that matter! She doesn't bother to think out a problem; just deals with the people involved. She bases her judgments on worth, evaluating things by emotionally toned experiences: I feel this is good, but that is bad. I feel this is morally right; that is morally wrong.[9] She twists the whole family around her little finger, but she's nice to have around when you need some understanding."

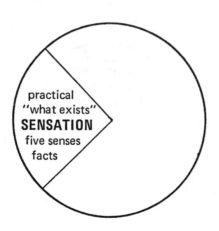

C. The group leader will then go to either side of his Superior Function on the wheel for a Primary Auxiliary Function, depending on which of these two Functions he uses more. He can use both of these as auxiliaries, but here again he will tend to use one more than the other since he can't be perceiving shadowy possibilities at the same time he is perceiving the concreteness of things. If he selects *Sensation*, he will again draw a circle and jot down the descriptive terms. He will explain that this is HIS Primary Auxiliary Function, for, next to Thinking, he uses it the most. But his brother Harry uses the Sensation Function as his Superior Function.

"Harry perceives things as they are by what he can observe with his five senses. He knows things exist when he can see them. He is concerned only with facts, and the strength and pleasure of the sensation. He is painfully practical, concentrating on plain solid sense. And he really gets upset when a project takes an unexpected turn he can't understand! He claims he's a 'hedonist' and seems proud of it.[10] And he can sometimes have all the details of an event, yet miss the general all over picture.[11] "

D. The group leader will then go to the last Function, in this case the *Intuitive* Function. He draws the circle and puts in the terms when they are given. He thinks of his friend Hub when this Function is mentioned.

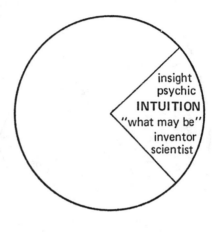

"When I ask Hub how he deals with a problem, I don't think he knows how he does it. He just plays a hunch and goes on from there. He seems to know the gist of the matter in an abstract way, and doesn't seem to be concerned with details or arriving at any reasonable conclusion or judgment. What makes it so hard to understand is that he is so often right—without taking the trouble to figure it out! That tends to bug me! He tells me he's going in for quantum physics. They're welcome to each other!"

Often it is hard to choose the Superior Function from between two Functions. If the player can figure out the Function he uses the least or represses, his Superior Function will be the opposite to this on the Function Wheel. This wheel can be found on the following page. The wheel also helps the player in determining his Function by showing him the types of people that combine the Functions (the in-between area) and the types of people that introvert and extrovert the Functions.

The Function Wheel [12]

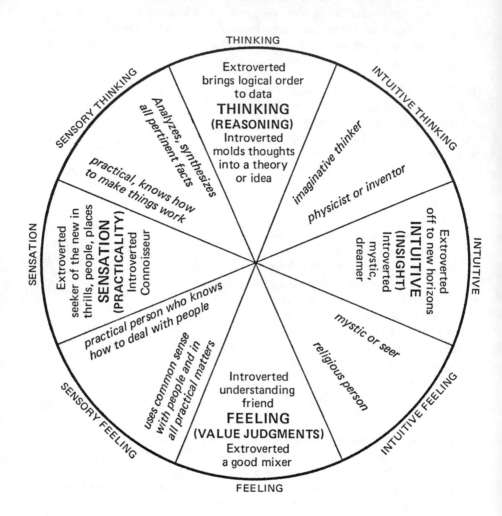

MANIPULATIVE BASKETBALL

Game 4

AIM: To gain an awareness of manipulation.

Everett Shostrom, author of the book, *Man, the Manipulator*, believes it is not necessary to reject manipulations—only to be aware of them—increasing awareness leads naturally to *actualization*.[13] He describes the term *actualizing* as creative behavior (Thou-thou) in contrast to the self-defeating behavior (It-it) of the person who manipulates.[14]

PLAYERS: Eleven players (five for each team) and a Referee
Four Forwards (2 on a team)—"Top-Dog" manipulators
Four Guards (2 on a team)—"Under-Dog" manipulators
Two Centers (1 on a team)—each "Taps" ball for his team
One Referee played by the conductor

The Center—The Center is the hardest position on the team and should be given to the best and sharpest players. The two Centers should be of equal fighting ability (or as near as possible). They will not only have to be well acquainted with the game, but know each player's Primary Manipulation and his opposite polarity.

Forwards and Guards—Each of these players will have to figure out his own primary manipulative pattern and choose this position on the team if possible. Try to have all eight types represented (they are listed below). But in any case, all types on one team must be different. Have two "Top-Dog" and two "Under-Dog" manipulators on each team; these will be the Forwards and Guards respectively. To help these players choose their positions and play them, Shostrom's types are listed below. The wheel on the following page will show "Top-Dog and "Under-Dog" positions and also the polarity of the type. Each player should know the polarity of his type; he will use this as his Reversal Role in the game. The actualizations of each manipulative position are also shown, with people who exemplify them.

Shostrom's Manipulative Types:[15*]

1. *The Dictator* exaggerates his strength. He dominates, orders, quotes authorities, and does anything that will control his victims. Variations of the Dictator are the Mother Superiors, Father Superiors, the Rank-Pullers, the Boss, the Junior Gods.

*From MAN, THE MANIPULATOR by Everett Shostrom. Copyright © 1967 by Abingdon Press. Used by permission.

63

2. *The Weakling* is usually the Dictator's victim, his polar opposite. The Weakling develops great skill in coping with the Dictator. He exaggerates his sensitivity. He forgets, doesn't hear, is passively silent. Variations of the Weakling are the Worrier, the "Stupid-Like-a Fox," the Giver-Upper, the Confused, the Withdrawer.

3. *The Calculator* exaggerates his control. He deceives, lies, and constantly tries to outwit and control other people. Variations of the Calculator are the High-Pressure Salesman, the Seducer, the Poker Player, the Con Artist, the Blackmailer, the Intellectualizer.

4. *The Clinging Vine* is the polar opposite of the Calculator. He exaggerates his dependency. He is the person who wants to be led, fooled, taken care of. He lets others do his work for him. Variations of the Clinging Vine are the Parasite, the Crier, the Perpetual Child, the Hypochondriac, the Attention Demander, the Helpless One.

5. *The Bully* exaggerates his aggression, cruelty, and unkindness. He controls by implied threats of some kind. He is the Humiliator, the Hater, the Tough Guy, the Threatener. The female variation is the Bitch or Nagger.

6. *The Nice Guy* exaggerates his caring, love, and kills with kindness. In one sense, he is much harder to cope with than the Bully. You can't fight a Nice Guy! Curiously, in a conflict with the Bully, Nice Guy almost always wins! Variations of the Nice Guy are the Pleaser, the Nonviolent One, the Nonoffender, the Noninvolved One, the Virtuous One, the Never-Ask-for-What-You-Want One, the Organization Man.

7. *The Judge* exaggerates his criticalness. He distrusts everybody and is blameful, resentful, slow to forgive. Variations of the Judge are the Know-It-All, the Blamer, the Deacon, the Resentment Collector, the Should-er, the Shamer, the Comparer, the Vindicator, the Convictor.

8. *The Protector* is the opposite of the Judge. He exaggerates his support and is nonjudgmental to a fault. He spoils others, is oversympathetic, and refuses to allow those he protects to stand up and grow up for themselves. Instead of caring for his own needs, he cares only for others' needs. Variations of the Protector are the Mother Hen, the Defender, the Embarrased-for-Others, the Fearful-for-Others, the Sufferer-for-Others, the Martyr, the Helper, the Unselfish One.

Preparation:

As in ordinary basketball, for this game you need to spend some time on skills and warming up sessions before you actually play.

Free Throw Practice—Have your players form a circle around the Referee-conductor. After giving an example of the Ball-Word, the Referee will explain what the Ball-Statement is and give an example to go with the Ball-Word.

The Shostrom Wheel [16*]

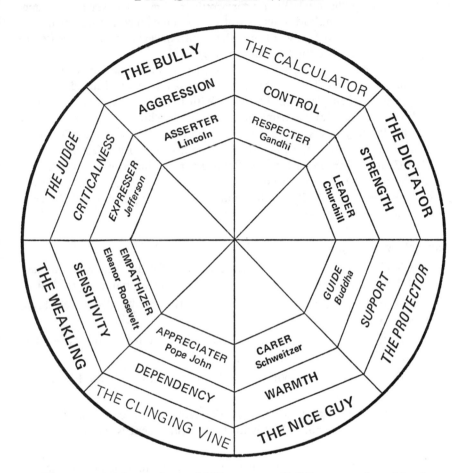

If anyone in the circle would like to try a Ball-Statement have him do so. If not, take the example and go around the circle, having each player give a manipulative reply from the easiest position (usually the Weakling or Bully). You may need a few examples to get started. For practicing (after the idea catches on), you could use a real basketball to throw to the player you want a reply from.

Center Skills—The people who will be the Centers should be given the chance to read over all the material, trying it out on their own until they get the feel of it. Until the group catches on to the game, the adult advisers might take the Center positions.

*From MAN, THE MANIPULATOR by Everett Shostrom. Copyright © 1967 by Abingdon Press. Used by permission.

Play Action of the Game:

Referee tosses up the "Ball-Word"—the word or phrase that starts the game. After hearing the word, both Centers will try to come up with a statement that is somehow related to the word and that will challenge their teammates to manipulate. This is called the "Ball-Statement."

The first Center to translate the Ball-Word into a Ball-Statement for one of his teammates wins the Toss. This Center is then said to have "Tapped the Ball" and the game is then played in his court. With the Ball-Statement this Center must also clearly call out the name of the teammate who is to respond, and if it is to be his Primary or his Reversal Role. If the designated teammate is able to respond correctly, then the team is given four points for a "Field Goal."

The Centers then go back to the Referee for another Toss-Up. When each player on one team has had a chance to respond with both his Primary Role and its opposite (his Reversal Role), the game will be declared over and the team with the most points is declared the winner.

Terminology and Rules:

Ball-Word—Word or phrase that the Referee gives out for the Centers to translate into Ball-Statements. The Ball-Word is heard by everyone in the game.

Ball-Statement—A statement loosely related to the Ball-Word that will bring forth a manipulative reply.

Primary Role—The manipulation that either the Forward or the Guard has chosen to play.

Reversal Role—The opposite polarity of the manipulative role the Forward or Guard has chosen to play.

Toss-Up—When the Referee calls out the Ball-Word, this is called the Toss-Up. The Center who responds first with the Ball-Statement is said to have won the Toss.

Tapping-the-Ball—When one Center gets the jump on his opponent by getting his statement in first, he Taps the Ball and the game is then played in his court.

Field Goal—When a Forward or Guard has successfully responded to the Ball-Statement, he has made a Field Goal which is good for *four points*.

Time! (referee)—Referee may call Time when he is undecided about any Ball-Statement or manipulative answer. He may get help from any of the players by questions. He will depend on each Center to defend his team's interests, and will let a good try go by unless the other Center objects.

Time! (center)—Center may also ask for Time, for he is the defender of his team. The Referee has many rules to remember and will depend on each

Center to be vigilant for his team's best interests. The Center will, however, only be alerting the Referee to a possible infringement. The Referee makes the final decision.

Free Throw—A chance for the opposing team to complete a Ball-Statement. Points received correspond to the points given for the Foul.

Fouls—Called when the Ball-Statement isn't brought home or if there is some infraction of the rules.

Two-Point Foul—This Foul is called when the Ball-Statement is not given a response. Sometimes the Center's statement is too far afield for the team-mate to rescue; sometimes the player may just not have a response. After a player gives up or a reasonable time elapses, a Foul is declared and a two-point Free Throw given. Any player on the opposing team can take this incompleted Ball-Statement and use his Primary Role to respond. If his reply is successful, his team will receive *TWO POINTS*.

One-Point Foul—(1) This Foul is used when one Center is repeatedly jumping in unprepared, i.e. with a Ball-Statement that is not related to the Ball-Word. If Time is called and the Center himself is unable to show the connection, the opposing Center is given a one-point Free Throw to attempt to change the offending Ball-Statement to comply with the rules. If the Referee decides he has succeeded, *one point* is awarded.

(2) The Center has a great deal of material to keep in his head and this is part of the game. If he forgets and calls for a Role that his teammate has already responded with, a Foul is called with a one-point Free Throw. A Forward on the opposing team may take the offending Ball-Statement and respond to it if he can, using his Primary Role. If he is successful, his team receives *ONE POINT*.

(3) If a Forward or a Guard uses an incorrect response or mixes his Primary and Reversal roles, he receives a Foul with a one-point Free Throw for the opposing team. A corresponding Forward or Guard on the opposing team may take the same Ball-Statement and try to answer it, using his Primary Role. If he is successful, his team receives *ONE POINT*.

Examples of the Play Action:

BALL-WORD: *Abortion.*

The Ball-Statement might be: "I'm going to have a baby!" The center will then call out the name of the player he wants to respond and which role he wants him or her to use. Example: Jack :: Protector.

Protector: "I don't care who's responsible, dear. I'll stand by. I can't bear the thought of your having an abortion. We'll get married. School and my career aren't important."

Other players could react to the same Ball-Statement:

Dictator: "It's about time someone took you in hand. I'll arrange for an abortion. I have influence."

Weakling: "Oh my goodness! What will you do? Have you been thinking of an abortion? This is too much for me. I'll worry myself into the grave!"

Calculator: "No big deal. Get an abortion. This would make interesting news to a few people I know!"

Clinging Vine: "What do you expect me to do? Oh dear, I'm not feeling well. I'm getting such a headache."

Bully: "Don't expect me to pay for any abortion. You got yourself into this, you get yourself out. This is just about what I would expect from *you*."

Nice Guy: "Oh you poor kid! I don't like to advise you on anything like this, but I'll help any way I can, you know that!"

Judge: "My dear! How could you do such a thing? Aren't you ashamed of yourself? An abortion may be needed."

A Ball-Statement like "I'm pregnant and you're responsible!" would be more directed to the Bully who could respond: "Why you silly dope, I was only one of a half dozen other guys! Don't try using my name, or you'll be sorry! Don't expect me to pay for an abortion!" Etc.

BALL-WORD: *Watergate.*

Teammates should help the Center "bring in" the Ball-Statement no matter how preposterous it seems. So if the Ball-Statement is: "The moon is made of green cheese," the Calculator as an intellectualizer could cite the moon expeditions and accuse him of lying to the uninformed; the weakling could just play dumb and withdraw by saying: "Oh, are you sure? I'm always puzzled by these things."

BALL-WORD: *Cheating.*

1. Ball-Statement: "Here is my exam paper, Teacher."
   Center to Forward: "Sally :: Dictator."
   The Mother Superior in the Dictator could respond: I hope it's all your own work, Bill. We must be careful to be honest if we hope to succeed in life."

2. Ball-Statement: "My three aces win the pot. Anyone want to make something of that?"
   a. Center to Guard: "George :: Weakling."
   The Giver-Upper with a touch of Stupid-Like-a-Fox: "No, of course not. I guess I'm just stupid. I was trying for aces, too."
   b. Center to Guard: "Harry :: Clinging Vine."
   Bringing out the Crier, the helpless One: "Oh, I'm wiped out. What will I do? Could you possibly spare some of the money for my children?"

3. Ball-Statement: "I'm going to be busy at work tonight so I won't be able to keep our date."

Center to Guard: "Debbie :: Clinging Vine."

Bringing out the Parasite in the Clinging Vine: "Oh, I had *so* counted on your being with me tonight. I get so lonesome sometimes I think I'll end it all."

Clarification of Rules and Helpful Hints

1. Exaggerate manipulations so there can be no question of their passing. Manipulation is the aim of the game.

2. Because everyone has heard the Ball-Word, it is not necessary for the Ball-Statement to make it clear. With the Ball-Word in mind the player should be trying to respond to what the Center has given in any way he can.

3. Centers are allowed some leniency. If a Center forgets the Primary or Reversal Role of a player, he can just call out the player's name with "Primary" or "Reversal," hoping his teammate can use it with his statement.

4. Ball-Words and Ball-Statements can imply different things and can be used in different ways. A teammate may use the Ball-Statement in a manner completely different from what the Center had in mind. It would still be correct.

5. It is not necessary to mention the Ball-Word in the manipulative answer to the Ball-Statement. But encourage the players to do so where they can, since it helps clarify their reasoning.

6. Forwards and Guards should take note: If you make no attempt to answer a Ball-Statement, the opposing team gets TWO points and any member of the opposing team can try to reply. This is a stiff penalty! It is a great deal stiffer than having your try judged a wrong response and losing only ONE point. Moreover, if the Referee feels it is a good try, he may say nothing. If the opposing Center does not object, it will get by. The game is structured to encourage trying.

7. Remember, the main purpose of the game is to get the Forwards and Guards to manipulate. If you are having a hard time getting anyone to fill the Center spots (because they are difficult), subsidize this position by allowing the Centers to see the full list from which you will be choosing the Ball-Word. In this way they can prepare a general Ball-Statement for each of the words. If using Primary and Reversal Roles makes it too tough at first, use only Primary Roles.

Suggested Ball-Words:

Watergate	Lying	Illness	Love
Open Marriage	Cheating	Blackmail	Sex
Threatening	Abortion	Comparison	Women's Liberation
Jealousy	Siblings	Prejudice	

Contrasting as a Discussion Technique for Manipulation[17]

THE MANIPULATOR (It-It Relationship)	THE ACTUALIZOR (Thou-Thou Relationship)

Deception

(Phoniness, Knavery)

a) Uses tricks, techniques.

b) Plays a role only to create a desired impression. Puts on an act to maneuver his antagonist.

Honesty

(Genuineness, Authenticity)

a) Is able to be himself, whatever it may be.

b) Expresses himself the way he is feeling it. Is candid and genuine; always leveling.

Unawareness

(Deadness, Boredom)

a) Has "tunnel vision," seeing and hearing only what he wishes to.

b) Unaware of the really important concerns of living.

c) Sees life as a battle.

d) Feels unappreciated and unvaluable, regards others the same way (as "its").

Awareness

(Responsiveness, Aliveness, Interest)

a) Looks at and listens to himself and others.

b) Fully aware of all the dimensions of living.

c) Sees life as a growth process.

d) Appreciates all aspects of himself and others.

Control

(Closed, Deliberate)

a) Regards life as a game of chess; calculates every move.

b) Concealing his motives, he controls himself and others.

c) Demands or is submissive.

d) Chooses to control others.

Freedom

(Spontaneity, Openness)

a) Is spontaneous.

b) Allows himself and others freedom to express potentials.

c) Not a puppet or an object, but master of his life.

d) Chooses to contact others.

Cynicism

(Distrust)

a) Distrusts himself and others; doesn't trust human nature.

Trust

(Faith, Belief)

a) Has deep trust in himself and others, and can cope with life in the here and now.

(Discussion with Contrasting is different from other Listings. Here you are trying to coax the meaning of "actualizor" from your participants. They will tend to remember the meaning better if they have thought it through on their own. Ask your players if they can name a trait of manipulators. When they have come up with one, write it on the blackboard, and then ask them to name the opposite trait, which you can write under *Actualizor*. Then ask them to define these traits; and write the opposing definitions under each. Proceed with traits and definitions one by one, letting your *players* do the thinking.)

Suggested reading:

Maria F. Mahoney, *The Meaning in Dreams and Dreaming* (New York: Citadel Press, 1966);

P. W. Martin, *Experiment in Depth* (Boston: Routledge & Kegan Paul, 1976);

Everett L. Shostrom, *Man, the Manipulator* (Nashville: Abingdon Press, 1967).

5

/earch for identity

A teenager with his fertile mind is like a block of clay waiting to be molded into a figure. The rough outline is there but more work is needed to determine what the figure will be. In understanding himself or herself (an essential part in the search for identity), a youth is taking this lump of clay and is starting to bring out a figure. The last chapter was designed to help each youth realize that he or she is a unique person, and take pride in this, in order to establish an invulnerable core as an individual. In this chapter, youths will go beyond finding out what they are and find out what they can become.

The emerging figure needs to become more distinct. How can the "givens" of the youth, his or her talents, abilities, and interests, be used? They are an important part of this figure. In all things, *mind is the builder*. Any youth can create his or her own path. With aims and goals in mind, the youth can bring about an exciting and fulfilling life. The desire and will of the individual will change these aims many times during a life time, but each time they will have an important part—in molding the clay.

THE BUILDING OF AN EGO

Roleplay 8   (disruptive type of role)

*Jocko: (*Trickster*—loves to play tricks) A person who loves to shock people and play tricks on them. He has found he gets attention this way, and at his stage of development he seems to need it. The conductor will encourage him to interrupt whenever anything constructive is going on. Usually when this is carried to the extreme it will get under the skin of the rest and they will register strong disapproval.

Harvey: (*Hare*—carries flag for a cause) He is easily "turned on" by a cause. An excellent salesman, he has sold most of the tickets for a raffle to send representatives to the Scout Jamboree. The conductor will encourage him to gain the support of his fellow Explorers for his cause.

Horner: (*Red Horn*—champion of the underdog) He seems to get a lift from helping his brother who is down. The conductor will advise him to defend Jocko but not until some disapproval is shown by the others. He will then laugh at Jocko's jokes and interruptions and identify with him.

Mat: (*Twins*—mature hero) He combines the forceful leadership of the extrovert with the considerate understanding of the introvert. The conductor should apprise him of the whole situation and what is being attempted. Mat may have to start the disapproval of Jocko's actions if no one else does. If the role gets too disruptive Mat may have to try to bring order out of chaos.

[The scene opens as Mat comes into the meeting room of the Explorers. Jocko has been tripping everyone as they come in the door. Mat has just managed to keep his footing.]

Mat: Cut it out, Jocko! Your attention everybody. Our leader won't be here until later and he's asked me to take charge. We might as well get started with old business.

Harvey: How about ticket selling for the Scout Jamboree? We're way behind our goal. I've got an idea that might help . . . .

Jocko: Well, what d'you know. The man has an idea.

Harvey: I've thought of something to make selling more fun . . . .

Jocko: Fun man, I'm all for it!

---

*Role-descriptions and dialogue are for the conductor only; they are given as guides.

Suggestions for the Roleplay "The Building of an Ego"

Selecting Your Players. Careful selection of players is necessary for this role. Try to select players who will feel comfortable in their roles and who will enjoy playing them. The players will be portraying the four stages of ego development of the Winnebago Indians.[1] Fighting ability is not as important here as identification with the role.

Instructing Your Players. In this role you will be attempting a *disruptive* type of roleplaying, so give specific advice about it to each of your players, individually, in front of your audience. The stages of ego development mentioned are only for the conductor. Avoid using titles, especially Trickster or Hero.

In giving your instructions, try to instill confidence in each player for his particular role at the same time that you are intensifying the conflict. To Jocko you might say something like: "Things are really dull around these Explorers' meetings and it's up to you to put a little life into them. You can't let them get too serious and turn into another school room! You can always find a trick or gag to pull whenever things are slow. Horner is a friend and likes a good time. Maybe he'd like to help?" And to bring him into the scene: "You're at the weekly Explorer meeting and ole sober-sides Mat has just walked in. How quick is this guy on his feet?"

To Harvey you might say: "The big raffle drawing to get money for delegates to the Scout Jamboree is only about a week away and there aren't nearly enough tickets sold. How can you get the guys into the mood for selling more? Will the prize your father offered (under pressure it's true!) stimulate them enough or will you need something more?" To make Harvey a part of the scene: "You've just gotten to the meeting and been told by one of the guys that the adviser won't be here until later. Mat will be in charge. Will he be able to help you? Will that silly ass Jocko disrupt the meeting again so that you can't get anything done?"

To Horner you could give something like: "You hope the fellows don't pick on Jocko again tonight. Jocko is a great guy, and just likes a little fun once in a while. Things would get really dull without him around." Etc. And to bring him into the scene: "You can see by the merry look in Jocko's eyes that he's in fine form tonight. You wonder where the adviser is. Will Mat know?"

To Mat you might say: "Your role will be essentially a prop role; it will be up to you to keep things rolling, and prevent chaos. You will handle Jocko's tricks in as mature a way as you can. If the others don't object to his tricks and interruptions, you may have to instigate some objections and curb him—getting the others to follow you. You will help Harvey in promoting the ticket sale, but wait for him to take the initiative." To plunge Mat into the scene, say something like: "You have just arrived at the Explorers' meeting and will have to get it started, since the adviser won't be here until later. Will you be able to manage the guys?"

What to Watch for. Watch to see how Jocko manages his role. Are his tricks subtle and a certain amount of fun, or are they crude and obnoxious? If they are crude, do the other players object, speaking out in a candid way, or do they hold their feelings inside in resentment? Does this resentment come out in put-downs and sarcasm? Or if the other players are enjoying Jocko, do they object to Mat's trying to stop him? Each way will present a problem for someone to work out.

Watch to see how Harvey gets the attention of the other players. How successful is he in interesting them in the sales competition? Does he persuade smoothly or in a haranguing manner? Has he made his cause their cause? Is he able to take Jocko's interruptions in stride?"

Take notice of how Horner helps Jocko. Does he help the other players enjoy the fun of "goofing off," or does he seem to be just another Jocko making everyone uncomfortable?

The role of Mat is a prop role, but how does the player handle it? Can he fend off Jocko's tricks and interruptions in a firm way but not belabor them? Does he manage to keep things rolling? If Jocko is popular, is Mat able to gain the support of his fellow Explorers in trying to stop Jocko? Has he been able to help Harvey interest the others in ticket selling? Does leveling or manipulation play a big part during the roleplaying?

Discussion Following the Roleplay. Explain to the Jocko-player that rejection of his tricks was part of the role. Find out how he was feeling during the playing. Ask the audience if they can see any parallel between the Jocko character and the Br'er Rabbit stories. Discuss any connection they might see between the part of Jocko and frustrated rioters' breaking windows and destroying things. How does Jocko compare with the natural child of the PAC concept of Berne and Harris (pp. 20, 98)? Discuss clowns, jesters, and comedians, who with exaggerations help us look at ourselves and laugh at our own imperfections. Bring out in the discussion how the natural child adds spice and fun to living.

How has the Harvey-player handled getting the attention of his fellow players? If he was not too successful ask the other players or the audience if they can think of better ways it might have been done. Ask the audience if they can see any parallel between the Harvey-character and the college students who worked for civil rights and other causes. If the Horner-player has been drawn down to Jocko's level of tricks, try to bring out in discussion how this may have happened. Bring your audience into the discussion by asking how they might have handled the problem.

Unless Mat has been a complete disaster, commend him for playing a difficult part. Have him tell what was the hardest thing he had to do and why. Find out if he was aware of any lack of leveling among the players. Did the Mat-player or any of the others use forms of manipulation?

"What Is Maturity?" as a listing technique (p. 104) can be used to advantage here. The game PAC Pinball (p. 98) is also applicable.

FORCES CRIPPLING THE EGO

Roleplay 9

*Terry: A teenage ADC mother with a baby. Life has dealt Terry one bad deal after another. Her father deserted her and her mother when she was a small child. Rather than get help from the government, her mother decided to work, and this left little time for Terry or housework. Other mothers supplied their daughters with the pill and made sure they took them. But not *her* mother! Her mother told her that if she was just going to lie around she'd have to start earning her *own* money. Then she got pregnant, and, not having the money, she had waited too long to have an abortion. Even then she had thought it would be nice to have a home of her own—and a baby of her very own to love her! But it hadn't turned out that way. Life now had become a drag and she wondered if it was worth living. If she had a lot of money it would change everything.

Gwen: A girl living in the apartment next to Terry's. Having just graduated from high school, she is working this summer to earn money to attend the university in the fall. She will be getting a degree in psychology—she wants to be a social worker. Although not a close friend, Gwen had known Terry in high school before she had her baby. Because Terry has been trying to cling to her in a one-sided friendship, Gwen feels she should jar her into leading a more balanced life.

[Action takes place in Gwen's apartment, in a building in one of the poorer sections of a university town. The scene opens with a knock on Gwen's door; she opens it and there stands Terry, slouching.]

(No dialogue is given here; as a guideline, the conductor can use the dialogue of the Open-End Play [p. 80].)

*Role descriptions are for the conductor only.

Suggestions for the Roleplay "Forces Crippling the Ego"

Selecting Your Players. With this role ask for volunteers. You might ask specific people if they would be willing to play Terry as a prop role. If no one wants to, it might be best to use the Open-End Play. In asking for volunteers to play the ADC mother, you should explain that it is a prop role of a girl who has had a rough deal in life and is depressed and discouraged. You will be trying to give the Gwen-player a foil to work against, but you will also be providing a medium for someone who can identify with the discouraged-girl part (do not make the latter goal too obvious, however). The players' abilities here are not as important as their identification with the roles.

Instructions for Players. Remember, the role-descriptions and dialogue (from the Open-End Play) should not be given to roleplayers; it will inhibit their creativity. Inspire your players for their roles by a confidential one-to-one talk. If you yourself can feel some sympathy for the part, this will carry over in your voice and "body language." To the Terry-player you might say: "Life has sure dealt you a bad hand. Maybe you ought to throw it in—simply pull out of this game. After all, is it worth so much pain? Some people have all the luck; you sure haven't been given any! Your father ran away. Your mother . . . ." Be creative and help your player be free to level about how she is really feeling. Bring her into the scene with something like: "You have just realized you need some groceries if you and the baby are going to eat today. You have absolutely no desire to go out. Maybe Gwen will get some things for you. She is usually pretty good about it. You go and knock on her door."

To the Gwen-player you could try something like: "You're going to try to get a degree in psychology so you can be a social worker. You've wondered what kind of help you will be, and Terry with her baby is an opportunity to find out. She is obviously discouraged and not likely to change with the way she is looking at life. Maybe you can help her." Etc. Bring your player into the role with: "There is a knock on the door. You go to answer it and there stands Terry. This is unusual; she is usually watching TV at this hour."

What to Watch for. Watch for ways the Gwen-player tries to help. Is she so sympathetic that she encourages Terry in her sullen approach to life? Or is she so rough on her that her help comes across as put-downs and makes the girl feel more depressed and bitter? Does the Gwen-player seem like a critical and chastising parent? Are the players leveling so that the action moves forward? Watch for defensiveness and resentment in the Terry-player, or manipulative helplessness. This is one role in which the Terry-player may be prudent to ignore what is uncomfortable for it might indicate a basically different philosophy. Such brushing aside, however, may not help the progress of the role. It should be noted so it can be brought out in the discussion period. Be careful not to condemn a player, who in spite of her real feelings, played the role as written.

Watch for manipulation. This can be from both the Gwen-player and the Terry-player. If a player does try it, how does the other one handle it? With more manipulation?

Discussion Following the Roleplay. If your Terry-player begins to feel anxious or upset, proceed with caution. Discuss the ways the Gwen-player tries to help. Are they effective? Ask the Terry-player how she felt about the help. Can the audience suggest other ways of helping or advice that may have been more effective? Have the audience list the crippling and positive points illustrated by the Open-End Play (as on p. 82). Help bring out these points by asking questions. Center a discussion on the points listed. How are they crippling? How do the positive aims tend to help?

If manipulation was used during the roleplay, bring it out in discussion. The conductor may wish to use the contrasting on manipulation (p. 70) as a discussion technique. If there was a great deal of manipulation, and the group is interested or curious, play the game "Manipulative Basketball" (p. 63). The game "The Royal Order of the Symbol" (p. 92) might also be played after this roleplay to emphasize how positive aims can change the direction in one's life.

FORCES CRIPPLING THE EGO

Open-End Play:

[This is an Open-End Play covering the same material as the above roleplay. The skit, showing the interaction of two people with different philosophies of life, will bring up points helpful to a discussion of a difficult subject. The scene opens as in the roleplay, when Gwen, the girl aspiring to be a social worker, opens her apartment door to find Terry, the teenage ADC mother.]

*Gwen: Come in, Terry, Your TV hasn't broken down, has it?

Terry: No. The news is on now. Martha said she thought you were going grocery shopping this morning.

Gwen: Yes, I planned to go about ten. Would you like to go along?

Terry: Nau. I just thought maybe you would get me a few things?

Gwen: It would do you good to get out, Terry. Keeping your eyes glued to that TV set is demoralizing.

Terry: It's the only decent thing in this crummy world! For a little while I can pretend I'm someone else.

Gwen: But what happens when the program ends and you're you again?

Terry: [shrugs shoulder] There's always another program. [looking around] It really looks nice in here.

Gwen: Your apartment's the same as mine—or was!

Terry: You haven't got holes in your plaster. I told the old landlord about it ages ago but he hasn't done anything. He makes me furious!

Gwen: Is that where your babysitter ran his bike into the wall?

Terry: Yeah. And he's knocked out other pieces since. That damn kid! I pay him to watch the baby so I can sleep and get a little peace. And, you know, sometimes he's more trouble than the baby.

Gwen: Maybe you had better fix the plaster yourself—and clean up the apartment. I have a feeling the landlord will move you out before he does anything there.

Terry: Oh, we'll just get moved into that new building they're putting up.

Gwen: Think you could keep that place up any better?

---

*This Open-End Play is to be used after (or in place of) the roleplay so as not to inhibit the creativity of the roleplayers.

Terry: Gosh, Gwen. [puzzled] I didn't think you would mind my having a nice place to live.

Gwen: I'm sorry, Terry. I'm just trying to show you that moving into a new place isn't the answer to your problem.

Terry: I didn't know I had a problem.

Gwen: I realize that. When you worked in the laundry, Terry, did you ever feel proud of a shirt you had ironed?

Terry: Heck no! I thought of how much I hated it and the big baboon who was our boss.

Gwen: Have you ever wanted to do some special work, like—like maybe fixing hair?

Terry: [walking toward mirror] My hair looks sexy when I fix it, huh? [looks in mirror and makes a face] Well, I haven't had a date in two months.

Gwen: Your hair is beautiful, and especially when it's freshly washed!

[Phone rings in Terry's apartment and she runs offstage to answer it. She can be heard saying 'Oh yes!' in an excited tone. She comes back into the room in a thoughtful manner.]

Terry: That was my friend Mary. She tells me there's a real slick guy looking for a date next Friday. And, well . . .

Gwen: What?

Terry: My place is such a *mess*! I wondered if maybe we could exchange apartments for the evening.

Gwen: So you wouldn't have to clean your own?

Terry: But there's the broken plaster and the spots on the rugs! The baby throws food all over and it gets tracked on the rugs. If I clean it, it'll be a mess in a short time anyway.

Gwen: And it'll continue to be messed until you start cleaning it up when the food spills.

Terry: What about the apartment?

Gwen: The answer is a definite no!

Terry: Hey, you aren't one of those people who are against an ADC mother getting a little sex, are you?

Gwen: No, Terry. In view of the rest of your life it probably looks like a positive element to you. But why not look for something more than sex?

Terry: Nothing wrong with sex, as far as I'm concerned. I don't like to get too involved. Damn guys expect too much. Then when you're hooked they desert you.

81

Gwen: Not all men are like your father, Terry.

Terry: Says who? Look, let's drop that. How about me fixing your hair? Will you switch then?

Gwen: Nope. Wait. [hesitates] You like fixing hair and you're good at it. I'll make a deal with you.

Terry: What kind of a deal?

Gwen: You fix my hair and clean your apartment, and I'll help you fix it up so you won't be ashamed of it.

Terry: The spots? And the plaster?

Gwen: I'll rent a shampooer, and if all the spots don't come out we'll use an area rug. I've used the plaster they use for drywalls and I can do a passable job. We can get some paint from the landlord. How about it?

Terry: Well . . . . I'll have to think about it.

Listing as a Discussion Technique: Forces Crippling the Ego

[After the Open-End Play the conductor may wish to have the audience list the different elements that are crippling. Then list what could be positive elements.]

*Crippling Forces:	*Positive Forces:
Asking for life as a gift.	Choosing work she likes.
Projecting own faults unto others.	Working with talents she has.
Non-producing.	Producing something creative.
Living against life instead of for it.	Accepting new ideas and trying them.
Retreating from life.	Being proud of something.
Contriving a life of imagery instead of real life.	Making realistic plans or aims for the future.
Using up energy in anger and hatred.	Making meaningful relationships.
Toying with a fairy dream of what she would like to be instead of what she can be.	Not letting fears of the past rob her of new experiences.
Avoiding deep relationships.	
Looking for gratification of feelings and entertainment rather than something deeper.	

*These are only suggestions. Allow your audience the freedom to come up with their own ideas.

82

FORCES FOR A HEALTHY EGO

Roleplay 10

*Clint: Brilliant high school senior who enjoys schoolwork and is challenged by new ideas—especially in physics. Clint's parents died many years ago and he has been brought up by guardians. They were very impersonal and Clint was often lonely. Now, however, he is on his own with full control of the money his parents left him. Up until now, Clint's inability to relate to people has been unimportant. The lab and his future career as a physicist were all that mattered. But now there is something new—something besides the money. Now there is Nancy!

Nancy: An attractive senior, who is determined to save Clint from the miserable life of being a slave to the establishment. She has seen his need to open up and feel the warmth of human relationships. She dislikes Mr. Paxton, the physics teacher, for he keeps Clint far too busy to have time for social affairs.

Mr. Paxton: High school physics teacher. He sees that Clint has the makings of a brilliant physicist. Aware of Clint's naiveté with girls, he feels Nancy is a danger to Clint's career.

[As the scene opens, Mr. Paxton joins Clint in the physics lab, where Clint has just finished an important experiment.]

Mr. Paxton: How did the experiment come out?

Clint: Great! But I wasn't sure until the end.

Mr. Paxton: You've got a great gift, Clint. You're luckier than most fellows your age. You know what you want to do and you have the ability and the funds to do it. You can go as far as you want to.

Clint: I could wish for a little more experience with girls and more ease with people in general. Nancy has suggested we join a group of her friends at a hippie retreat in California this summer. She claims she'll teach me.

Mr. Paxton: With LSD?

Clint: Come now, Mr. Paxton, I'm not stupid.

Nancy: [coming into the room] Who's not stupid?

*Role-descriptions and dialogue are only for the conductor; they are given as guides.

83

Suggestions for the Roleplay "Forces for a Healthy Ego"

Selecting Your Players. Equal fighting ability would be good here but players' identification with their parts is more important. If you know your audience, select people you know will fight for each particular viewpoint. If not, ask for volunteers.

Instructing Your Players. In this role you will be trying to show how a healthy ego is a balance between positive forces. You should encourage each player to fight for his point of view. It will be up to them to come up with compromises. Try to feel some sympathy for each of the viewpoints; this will carry over when you give your instructions.

To the Clint-player you could say something like: "You have sometimes been lonely without parents or brothers or sisters. Nancy is a new experience—she certainly opens up new vistas of learning—but she also interferes with your work." Etc. And, to bring him into the scene: "You're glad that Nancy is late—you'd never have been able to finish with her there. You want to see her, but later. You're happy to see Mr. Paxton come in the door; you need to talk about the experiment."

To the teacher-player you could say: "You dislike this girl Nancy who is running away from life. She's a bad influence on Clint; except for her he could probably do brilliant things in physics. Clint's parents have left him enough money for whatever education he will need—*if* Nancy doesn't have him throw it away on one of those communes full of freeloaders!" And to bring him in: "Clint's in the lab and you're going to have a talk with him."

To the Nancy-player try something like: "Clint is a real challenge! You have never met anyone so deprived of human relationships. And to top it he's planning a career where he'll be working alone. He needs a job, for a while at least, where he'll have to work with people. You feel a farm or a commune where he'll be able to form new relationships is almost a necessity at his stage in life." Etc. And to plunge her into the scene: "You are late on purpose hoping Clint will be finished with his experiment. He doesn't seem human until he is. You walk into the physics lab, and Clint is there but so is that Mr. Paxton whom you dislike."

What to Watch for. How do the players go about convincing Clint of the validity of their points of view? Do they realize that talking to him alone, without a third person, can accomplish a great deal more? Watch for manipulation from both Nancy and the teacher-player. Are they giving Clint breathing room to make his own decisions? Or are they pressuring him? Are they using unfair tactics? Watch for leveling in a sincere tone that speeds the action or nonleveling that slows it down. Are put-downs or sarcasm being used? How is the Clint-player reacting to this? How is the Clint-player handling his role? If some form of manipulation is used, how is the Clint-player reacting?

Discussion Following the Roleplay. Ask the Clint-player if he felt pressured by either of the other players. Did he feel free to make his own decision? Was he drawn to one player more than the other because of the way that one presented facts? If put-downs or sarcasm were used, was the Clint-player influenced against the person using them? Discuss how the Nancy-player and the teacher-player tried to convince Clint of their points of view. Was each willing to compromise? If leveling or nonleveling was evident, bring out the difference it made. Ask the audience's reaction to this. How did the Clint-player handle his role? Did he manage to listen to one player without antagonizing the other? If neither of the other players thought of asking to talk to him alone, did the Clint-player suggest to one or the other that he would like to do this? Was he able to do it in a gentle way without hurting or antagonizing?

Don't forget the games at the end of this chapter and also the games on understanding (pp. 58 and 63); they can aid in the search for a well-balanced personality.

FREE TO BE ONESELF

Roleplay 11

*Sally: At sixteen Sally has already received a high school degree. Her parents felt she was too young to attend college, so Sally had found a job to earn the money to continue her education. She has insisted on living with her sister in an apartment along with some other girls. She loves her parents but realizes they tend to prevent her from growing up.

Mr. Brown: Father of the girl. He resents his daughter's moving into an apartment instead of saving her money by living at home. He feels she does not appreciate the value of money. He is hurt that she would want to live away from home.

Mrs. Brown: Mother of the girl. She does not believe her daughter is mature enough to be on her own. She misses her baby!

[The scene opens in the family car, where the mother has been waiting. Father and daughter have just come from the police station where the father has paid his daughter's fine so that she will not have to spend the night in jail.]

Sally: Thanks, Dad. I'm sorry to bother you but there didn't seem to be anything else to do.

Father: That's what a father is for.

Sally: Thanks just the same. It's hard to believe, but we had no part in the rioting; we were just going to our apartment.

Mother: Police don't usually pick up anyone without a reason.

Sally: They said it was because we shouldn't have been on the street after the order to disperse.

Father: Are you trying to tell me the police would arrest you for just going home?

Sally: Yes, Dad. That's exactly what I mean.

*Role-descriptions and dialogue are for the conductor only; they are given as guides.

86

Suggestions for the Roleplay "Free to be Oneself"

Selecting Your Players. This is a conflict role. Equal fighting abilities are desirable but identification with the roles is even more desirable—if you can find players who can identify. The parent roles here are to bring out the Parent Ego State (see p. 99) of your players. A girl with nurturing tendencies and a boy with an authoritarian parent will make the role move more easily; but they are not necessary. The true nature of your player's Ego State is often not discernible until after the roleplaying has started. Occasionally this will cause a reversal of roles and blocking-out of instructions. All this is part of what makes roleplaying interesting.

Instructing Your Players. This role is to show how preconceived ideas (fair or unfair) can hamper understanding. The role is based on a true incident. In a student housing area of a midwestern town, a riot occurred when police tried to stop an illegal block party. Many of the students (and even some adults) who had taken no part in the riot were arrested. Remember, the role-descriptions and dialogue given are for the conductor. They are given not just to set forth the facts but to help the conductor get his players into the mood and conflict of the role.

You could say something like this to the daughter-player: "You graduated from high school at sixteen and were perfectly capable of going on to college but your parents felt you were too young. Now you are holding a job capably and living on your own and they still treat you like a baby. You need to get away from their protective custody to be able to grow." Help her to step easily into the role by setting the scene: "Dad had paid your bail but had said nothing in the station. You had tried to jolly him out of a black mood but he wasn't responding. You follow him out of the station and see your mother in the car with that 'my poor baby' expression. She is opening the car door for you."

From your mother-player you should be trying to call forth the nurturing or protective instinct. Be creative. Convince the mother-player that her daughter is too immature to be on her own. Use examples from your own experience. It could run something like: "Your daughter is trying to fly before her wings are fully grown. You allowed her to be on her own because she was so insistent, but had it been wise? Are you being a responsible parent? Being allowed to eat an atrocious diet and ruin her clothes is one thing, but being arrested for rioting is something else again. Are you shirking your parental duties?" Bring your player into the scene with: "There she is, coming out of the station and laughing as if it's all a joke. She's such a baby! You open the car door and she gets in."

From the father-player you are trying to bring forth a sense of being hurt because his daughter prefers to live away from home. Say something like this: "Your daughter has no concept of money. She'll spend a dollar to get five dollars' worth or to get a dime's worth—it's all the same to her. She ought to be saving her salary for her education. But let Dad provide for that! Bring home the money, that's all kids expect from the old man these days!" Etc. And to plunge him into the role you might say: "You paid the bail and are now walking out to the car. You hadn't said anything in the police station in front of all those people, but now you're going to insist that your daughter come home. You open the car door on the driver's side."

What to Watch for. Watch how your parent-players pick up your instructions. Does the father-player play the role as an authoritarian parent? He will respond in the way he has learned from his parents. Does the mother-player play the nurturing parent or is she reasonable about her daughter's living on her own? Are they leveling about the facts and about their feelings? Is sensitive listening taking place? Do any of the players use active listening (i.e., giving sensitive feedback)? Is the daughter-player showing her need to be an independent person? Is she aware of the love and concern of the parent-players or only of her own need to be free?

Discussion after the Roleplay. Discuss what the young people feel the role of the parent should be. Is there a right and a wrong way of helping the developing teenager? How much freedom should a teenager be allowed? How many rights do the parents have? If a compromise was worked out in the roleplay, how does the audience feel about it? Discuss how the players used your instructions. Were they reasonable and honest about their feelings? Did the players' way of playing parents complicate the action? Bring up where leveling speeded up the action and where nonleveling slowed things down. Were any of the players left with a hurt or unsatisfied feeling? If any parallel speaking (see p. 17) took place, discuss it.

UNFAIR LOSS OR UNDESERVED HONOR

Roleplay 12

*Koren: A ninth grader, beginning her last year in junior high. Koren always checks the bulletin board before leaving for home. Officers for the Girls' Athletic Association will be posted, and she expects to be chosen president. She glances at the board and cannot believe her eyes. *Darlene* is listed as president of both the student council and the GAA! Koren's name does not even appear as a lesser officer. She is by far the best girl athlete, even if she isn't a top scholar. This isn't fair! She sees Darlene coming down the hall and hurries away. She has no desire to talk to her or anyone. Darlene is popular with the students and teachers, but that doesn't mean she's the best athlete!

Darlene: Also a junior high senior, Darlene has just found out that she's been chosen council president by the students. She is expecting her best friend, Koren, to be named president of the GAA, since Koren is by far the best athlete. They can celebrate together. As Darlene approaches the bulletin board, she sees Koren and calls out to her. Koren gives her an angry look before taking off at a run. Darlene looks at the board and finds her name listed as president of both groups; Koren is not even mentioned.

[Darlene hurries after Koren and manages to catch up. Although Koren will not even look at her, Darlene feels she must talk.]

Darlene: Gosh, Koren, don't be sore at me just because some of the teachers are stupid.

Koren: I'd rather not talk about it.

Darlene: Okay. I can understand how you feel. I just wanted you to know how I feel. I wasn't even expecting to be chosen as an officer of GAA, much less president.

Koren: But it's nice to be chosen isn't it?

Darlene: Yes, I'm human. But I feel better about being elected to the council by students than picked by teachers for the GAA. I'm a good athlete, but you're outstanding! I suppose it seems unfair to you.

Koren: You bet it does!

Darlene: I feel uncomfortable. Perhaps we could talk to one of the teachers.

Koren: Teacher's pet would think of that!

Darlene: They might give an explanation. I really can't understand it.

*Role-descriptions and dialogue are for the conductor only; they are given as guides.

Suggestions for the Roleplay "Unfair Loss or Undeserved Honor"

Selecting Your Players. Your players should be of equal fighting ability. The role will be more interesting if you can find a player (or players) who has (have) had an experience like the one here. If you don't know whether any in your audience have had such experiences, you might ask for volunteers.

Instructing Your Players. You are trying to instill confidence in each player about the role she is to play, and your tone of voice and "body language" will help you do this. You should also try to heighten the conflict. To Koren you might say: "You won't ever want Darlene as a friend again. You can't stand the sight of her. She has the position that is rightfully yours. Everyone knows this and you hate her! It isn't fair! It just isn't fair!" Etc. Remember to bring her into the scene with something like: "You see Darlene coming and you leave. You don't want to talk to her. She runs and catches up to you. That doesn't mean you have to talk to her. You suppose she feels very smart being chosen president of the council *and* the GAA."

To Darlene you could say: "Why is Koren sore at *you*? You didn't expect to be president. It made you feel uncomfortable. You two have been close friends since childhood. Can't she at least talk to you about it? Isn't she being unfair?" Etc. And to plunge her into the scene: "Once out of the building, you run to catch up with Koren. She won't talk and looks as if she is sulking. How childish can she be?"

Be creative about your instructions. Add what you can to increase the conflict or to help a player. Try to keep your help equal or to give more help to a lightweight player. Remember you are trying to bring out what it feels like to be treated unfairly or to win when you don't deserve to. Encourage your players to play their roles the way they are feeling them.

What to Watch for. Everyone will lose out many times during a lifetime—and many of these losses will be unfair. Both undeserved honors and losses bring on feelings that are difficult to handle and often hidden. This roleplay will give your players a chance to expose bad feelings, without putting anyone under actual pressure. Leveling is important here, but so is the ability to release feelings without antagonizing. The dialogue given shows how leveling helps get at the problem almost immediately. If you have sparked a real conflict or triggered some experience, your players may be kept from leveling by their resentments. Resentment may also prevent sensitive listening. In the role-dialogue given here the sarcastic remarks indicate resentment. Watch for this type of remark, as well as put-downs and overstatements, from the player with the unfair loss. Do they indicate resentment? Are they showing hurt feelings or some other hidden message? Is the resentment in the player with the unfair loss causing her to close off any genuine concern from her opponent? Is the undeserving winner using sensitive listening to be aware of the reason behind such remarks? Or is she thinking only of her own feelings?

Discussion Following the Roleplay. Discussion should be started by giving both players in the cast a chance to tell how they felt and why they reacted as they did (if they can). Did either of them recall an actual experience? If so, did the player or players level about the emotion they were feeling? If not, can they do so in the discussion? Was either player judgmental? Were "You-Messages" (see p. 18) used rather than "I-Messages"? What did the audience think of the way the players reacted? Discuss resentment. How it colors what we say, and why it makes it hard to level. Discuss the type of sensitive listening that hears hurt and resentment behind sarcastic remarks and overstatements. Does knowing the hidden reason for such remarks make it easier to avoid retaliating and help deal with them in a gentle way?

THE ROYAL ORDER OF THE SYMBOL

Game 5

AIM: To inspire ideals and goals.

MATERIALS: A sword, photocopies of the names for each player.

PEP TALK: (Give orally using own creativeness to add to it.)

"Life can be just as interesting and exciting as we want it to be. To be this way, however, it must have aim and direction, and it must be individualized for each one of us as persons. Everyone has certain 'givens' and 'talents,' but it's up to us as individuals to look within ourselves to find them and develop them. A good way to do this is to choose ideals and set goals to work for. In the game, you will be picking a name as a symbol to reflect the ideal or goal that you choose."

Game Action:

As a place to start ask your participants to examine their "givens," their interests, talents, and abilities. How can they use them to make a more exciting life? What profession, hobby, occupation, or avocation will use these "givens" to advantage? What skills, characteristics, or even virtues will be needed to carry this through? Have they need of some special trait to give them self-confidence or make their goal possible? What do they need to get the most out of life? Give each of your participants a list of names and ask them to pick a name that will reflect their interests in some way. This can be either as a profession or work, or a trait they believe advantageous to acquire.

Help Them to Think It Through in this Manner:*

1. Do you like people and like to be with them?
   You might enjoy being a *personnel director*, a *nurse*, or a *social worker*. If so, you would need to be *keen-eared*, *kind*, *merciful*, and a *comforter*. Or would you rather be a salesperson? Here you will need to be *cheerful*, *optimistic*, and perhaps *beguiling*.

2. Are you fascinated by a certain subject or profession?
   You might want to be an *actor* or *actress*, and would need *talent*, need to learn to *speak well* and be *vivacious*.

*Note: the italicized words can be found in the name-meaning lists (pp. 95-97).

Or maybe a *dancer* who would need to be *graceful, gifted,* and *lively.* Or perhaps be *exotic* or like a *flame.*

You might want to be an *ecologist* or *historian,* and would need to be *zealous* and *industrious.*

Or a *prophetess,* a *doctor,* or a *minister,* and would need to be *wise, helpful capable,* and *adept.*

3. Do you like action and adventure?

   You might like to be in the *coast guard,* a *Marine, Wac,* or *game hunter* and would need to be *brave* and *bold.*

4. Do you have *musical talent, craftmanship,* or *mechanical ability,* or an addiction to *writing?*

   To succeed here you will have to acquire *endurance,* become *resolute, unwavering,* and *patient.*

5. How about a political *leader* who would need to acquire *venerable wisdom, honesty,* and be *steady* and *righteous?*

6. Anyone for a *wife* who is *tender, lovable,* and *loving?* Or a mother who should be *gentle, serene,* and *motherly?*

7. Or a *husband* who is *helpful, courteous, faithful?* Or a *father* who is *kind* and *steady?*

   (Note suggestions in parentheses under occupation meanings [p. 96] and use them to stimulate players who have problems. If another sex will help for a specific meaning, change name to comply. Remember the name is only the symbol used, so if it becomes necessary create a new name.[2])

After names are chosen, pick a Grand Vizier, preferably someone who has chosen the category of a leader. He or she will find out the name, and the meaning of the name, of each of the players. Then the Grand Vizier will ask each player in turn to kneel and will touch each player's shoulders with a sword. She or he will dub each supplicant with the chosen name beginning with "Sir" for each boy and "Duchess" for each girl, giving a short spiel about the category of the chosen name.

Examples: 1. "I dub you 'Sir Wylie.' This gives me great pleasure for never have I met anyone so charming and beguiling."

2. "I dub you 'Duchess Thelma' for your wonderful nursing has saved many of my knights."

After everyone has had a turn, the Grand Vizier will say: "New knights and ladies of the Royal Order of the Symbol, you have been given a new name. It is up to you to see that the name performs its magic for you. You will spend a few seconds each day in affirming its purpose. Sit quietly and think of the name, declare its meaning positively, intellectualize it, feel it. Think of someone you admire who has this quality, or is in this position. Daydream of using it, of being caught up in it actively. Picture its accomplishment." The Grand Vizier will use his own name as an example:

1. "I would like to be a political leader so I chose the name Nestor which means 'venerable wisdom.' I will need to sense the underlying truth, and know what will be good for my country. Because of my strong desire for wisdom, I begin to feel wise on a deeper subconscious level. I think of Gandhi and Churchill. I try to picture myself like them. I see myself among a group of legislators trying to make a decision. I then picture myself hearing what has resulted from this decision, and I have a warm feeling of knowing I have done well." Or

2. "My name is Omar; I'm an architect. An architect plans a building, and works with people to construct it. I feel the joy of creating something beautiful. I see an architect I admire very much. I try to picture myself in his place. I can see myself working in close harmony with other people to bring these ideas to fruition. I can see majestic buildings I have designed." Or

3. "My name is Ada; I'm joyous. Joyous is being cheerful and loving life. I feel the sensation of joy permeate my being. I can see my friend Mary who is always bubbling with joy and I try to picture myself like her. I can see myself bringing joy to those around me. I daydream that the quality now has become a part of me, and my friends like to be near me so they can share my joy."

The Grand Vizier asks everyone to sit down and try it out for practice. Each will do it silently while the Vizier speaks:

"Sit comfortably! Quiet down! Think of your new name; its meaning. Elaborate on the meaning, finding other words to describe it." (Allow a few seconds to pass.) "Feel the emotion, the action, the being of it!" (Pause.) "Picture a person who has the quality you admire, or need, or is the best in the field you hope to be in. Picture yourself in that person's place. Daydream of a situation where you are using the quality, or are active in a project involving your name's meaning." (Pause.) "Now picture a time when you are at ease using the quality and picture also the pleasure it gives you. Or picture the work your hands or your brains have constructed.

"Remember that mind is the builder. As we seek, we find; as we knock, we are heard. What you can conceive, you can do. If your desire is strong enough, you can bring it into being. Positive thinking has a magic of its own."

Names to Inspire Boys

alert—Bryce
ardent—Ignatz
brave—Kimball
bright—Osbert
champion—Neil
cheerful—Tate
eager—Arden
famous—Elmer
fiery—Brant
free—Kermit
friend—Alvin
friendly—Elmo
foxy—Todd
gentle—Kevin
harmony—Alan
helper—Alexis
honest—Drew
kind—Holden
lively—Vivian
lucky—Felix
manly—Andrew
merry—Hilary
noble—Albert
order—Cosmo
rich—Otto
safe—Titus
sincere—Ernest
strong—Arthur
sunlike—Samson
wise—Conroy

a fighter—Boris
ambitious—Abelard
bold—Archibald
comforter—Nahum
courteous—Curtis
dextrous—Dexter
enduring—Durand
exceptional—Angus
faithful—Dillon
formidable—Egan
genuine—Sterling
growing—Vernon
honorable—Jarvis
industrious—Emmet
keen-eared—Otis
learner—Prentiss
lovable—Erastus
majestic—Augustus
merciful—Clement
outstanding—Jethro
patriotic—Leopold
peaceful—Culver
pleasant—Farand
powerful—Richard
renowned—Rodney
respected—Eldon
steadfast—Ethan
stronger—Xenos
vigilant—Gregory
valiant—Farrel

beguiling—Wylie
beloved—David
bright fame—Robert
capable, adept—Druce
charming—Wylie
clear one—Clarence
earthy—Adam
flame—Edan
heroic—Curran
illustrious—Clarence
kingly—Eric
laughter—Isaac
laurels—Lawrence
life—Hyman
lionlike—Llewelyn
longed for—Saul
mighty as a bear—Barrett
old in counsel—Eldred
pledge—Homer
renowned ruler—Roderick
shining light—Sinclair
shining of mind—Hubert
sophisticated—Desmond
steady—Hector
the greatest—Maximilian
unswerving—Hector
unwavering—Constantine
venerable wisdom—Nestor
watchful—Ira
young, virile—Colin

artist (painter)–Terrel
baker (caterer)–Baxter
brickmaker–Tyler
builder–Omar
   (contractor, architect)
chief–Cedric, Malvin
   (executive, magistrate,
   administrator, etc.)
chief, nobleman–Earl
   (federal cabinet officer)
cowman–Byron
craftsman–Wright
   (artisan, carpenter)
deep thinker–Edsel
gamekeeper–Warren
guide–Guy, Wyatt, Guyon
helper of men–Alexander
hunter–Falkner, Monte
   (detective, G man)
increaser (father)–Joseph
latheworker–Turner
   (machinist, die maker)
leader–Duke
liberator–Lysander
minstrel–Baird
one who summons–Sumner
oracle (prophet)–Phineas
peacemaker–Wilfred
poet (journalist)–Devin
roofmender–Thatcher
sailor–Murray, Murdock
scholar–Culbert
teacher–Enoch, Latimer
   (guru, trainer, coach)
warrior–Luther, Roger
worker in stone–Mason
   (sculptor, geologist)
world power–Donald

adviser–Redmond
   (business specialist, personnel
   director, lawyer, social worker)
arrow maker–Fletcher
candle maker–Chandler
chief, guardian–Howard
   (inspector, sheriff, fire chief)
chief of the valley–Kendall
   (mayor, governor, etc.)
coast defender (coastguardsman)–Seward
dove keeper–Coleman
farmer–Fabian, Barth, George
   (agronomist, agriculture professor)
guardian, warder--Parry
   (prison, port, or fire warden;
   caretaker; museum custodian)
healer–Asa, Galen, Jason
   (doctor, psychiatrist, medical
   technician, minister)
lover of the earth–Demetrius
   (environmentalist, ecologist)
people's ruler (president)–Theodoric
protector–Edmund, Richmond
   (policeman, fireman, ranger)
record keeper–Chauncey
   (accountant, registrar, scribe,
   bookkeeper, clerk, chronicler,
   genealogist, computer operator)
sea leader (commodore)–Marmaduke
speaker, interpreter–Driscoll
   (actor, preacher, commentator)
storekeeper (owner, clerk)–Spencer
   (grocery, drug, hardware, furniture,
   or department store)
wagonmaker (cars, trucks)–Wayne
war guardian (guardsman)–Hilliard
weaver (rugs, cloth)–Webster
   (manufacturer, designer, worker)

Names to Inspire Girls

beautiful—Linda	*ambitious*—Beverly	
dainty—Mignon	*beloved*—Darlene	
fertile—Pomona	*born free*—Camilla	
fiery—Brenda	*brave*—Bernadine	
flame—Bren	*brilliant*—Alberta	
free—Fanny	*cautious*—Prudence	
gentle—Mildred	*charming*—Dulcie	
gifted—Pandora	*cheerful*—Allegra	
graceful—Grace	*desirable*—Willa	
helper—Alexis	*faithful*—Fidelia	
helpmate—Sacha	*fervent*—Ernestine	
honest—Amena	*good luck*—Holly	
just—Justina	*gracious*—Roanna	
kind—Pamela	*happy*—Ida, Felice	
life—Eve, Vita	*hardworker*—Ilka	
lively—Vivian	*harmony*—Concordia	
lovable—Amanda	*honorable*—Noreen	
loving—Mabel	*hospitable*—Zenia	
lucky—Gada	*industrious*—Emily	
motherly—Rhea	*joyous*—Ada, Narda	
pleasant—Naomi	*logical*—Akilah	
pleasing—Hedy	*majestic*—Augusta	
rainbow—Iris	*married*—Beulah	
rebel—Asiyah	*merciful*—Mercedes	
reborn—Renee	*optimistic*—Nadine	
rejoice—Kay	*pledge*—Gize	
a song—Carmen	*promise*—Giselle	
speech—Amira	*rejuvenation*—Edna	
spring—Aviva	*resolute*—Constance	
a star—Esther	*righteous*—Camilla	
striving—Amelia	*serene*—Delphine	
strong—Irma	*speak well*—Eulalie	
tender—Morna	*spry sprite*—Disa	
truthful—Alice	*steady*—Pierrette	
untamed-Wilda	*talented*—Pandora	
wisdom—Ophelia	*vivacious*—Tallulah	
wise—Belinda	*well-known*—Lara	
zealous—Ardis	*womanly*—Charlotte	

Occupation, Avocation

adviser (lawyer?)—Monica
authority (boss?)—Hazel
battlemaid (Wac?)—Matilda
be successful—Barika
chief (manager?)—Melvina
commander—Hazel
composer—Edda
earth lover—Georgiana
friend of mankind—Sandra
gardenworker—Hortense
girl of forest—Silvia
glorious leader—Kim
God-consecrated—Isabel
harvester—Teresa
helper of men—Alexandra
historian—Clio
home mistress—Harriet
horse lover—Philippa
magic dancer—Satinka
messenger—Angela
mother—Ambika
musical—Cecilia
nursing—Thelma
one who heals—Emma
pattern (model)—Norma
peacemaker—Wilfreda
prophetess—Cassandra
protectress (cop)—Ramona
ruler (politico)—Roderica
scientific—Haley
sea maiden (Wave)—Alima
sea protector—Meredith
strong worker—Millicent
the greatest—Maxine
tower of strength—Magda
trader—Yarkona
watchwoman—Greer
weaver—Penelope

PAC PINBALL

Game 6

PLAYERS: The game is best played with from four to six players.

MATERIALS: Heavy cardboard (coat box); colored paper and paste; stapler or staple gun; bright flashlight; scotch tape, paper, and pencil. Noisemaker optional.

AIM OF THE GAME: To get your participants in touch with their Ego States. The game will be played by answering questions as honestly as possible. Explain to the players that the Ego States brought up in the game are not necessarily good or bad; all are necessary for a balanced life. The players will keep their own records, so encourage them to be honest. If they feel a guilty or negative feeling in regard to some of the questions, have them note where and bring it up in discussion.

Building the "Machine" (this can be done beforehand):

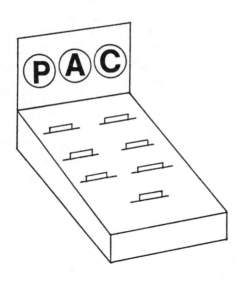

Take the bottom half of a cardboard coat box and cut off a small amount of each long side to give it a slight bevel. Cut part off one of the ends to match the level of the long sides. Invert the box keeping the short end in front, and cut in slits for the insertion of cards. For the headboard, cut a section out of the box's top; the headboard should stand up at least twice as high as the base. In the top of the headboard, cut three circles the size of a flashlight lens. Cover each hole with colored paper using a different color for each. Cut the letters *P*, *A*, and *C* out of black paper; paste one on each of the colored circles. Staple the completed headboard to the base.

(The machine is not absolutely necessary for the game. The conductor could just draw the circles on the blackboard and use colored chalk to outline a P, A, or C for emphasis. Or he could just express it verbally.)

Terminology of the Ego States (P, A, and C)

PARENT EGO STATE:[3] Taking on the attitudes and behavior of the people who have served as parent figures for the child.

> *nurturing parent*—protective, sympathetic, and supportive.
> *prejudicial parent*—critical, opinionated, chastising, and moralistic.

ADULT EGO STATE: "Feelings, attitudes and behavior patterns that are adapted to the current reality and are not affected by parental prejudices or archaic attitudes left over from childhood."[4]

CHILD EGO STATE:[5] Responding (to the inner world of feelings, experiences, and adaptations) in the same way you did as a child.

> *natural child*—Becoming the unrestrained infant still inside each person. (free-swinging, pleasure-loving, curious, sensual, impetuous, affectionate, but also rebellious, self-centered, and easily angered)

> *little professor*—Using the unschooled wisdom of the child, able to receive non-verbal messages and play hunches. (intuitive, creative, and manipulative)

> *adapted child*—Acting like a child who has modified natural inclinations to the demands of authority figures. (obedient, passive, anxious)

Play Action of the Game

Part 1. In this part of the game the players will be defining the area they wish to work in. Have them jot down the activities they engage in during a full year. Each player will choose ten items he or she is most interested in and spends the most time on. If a player spends a great deal of time on sports, her list might contain two or three different ones. If a player is a faithful watcher of TV, his list might have a number of programs. Another player's list might not have sports or TV at all.

> *Sports:* Football, Tennis, Swimming, Skiiing, Track, etc.
> *Lessons:* Musical Instruments, Singing, Dancing, etc.
> *Group Activities, Clubs:* Church Youth Groups, Debating, Choir, Cheerleading, Dramatics, Chemistry Club, etc.
> *Hobbies:* Photography, Coin or Stamp Collecting, etc.
> *Television* or *Reading:* News, Novels, Movies, TV Games, etc.

Part 2. As soon as one player is ready, start with him and allow the rest more thinking time. Of the ten items, have your player pick one that he spends a good deal of time with. Then start your game.

Have the player compute the amount of time he spends with this particular activity on a one to six point basis. Then have participant select one of the following answers:

I engage in this activity because:

1. I believe it's a sensible, logical thing to do. (A)
2. I think it is something I ought to do. (P)
3. I enjoy it—it's just for the pure joy of doing it (or watching it). (C)
4. My parents seem to like this sort of thing, and—it just seems the thing to do. (P)
5. My friends have joined and I enjoy being with them and doing what they are doing. (C)
6. It can teach me things that will probably help me later in life. (A)

(Explain to your participants that you realize their reason may not be listed here but ask them to choose the one that seems to apply the most or that seems to carry the most weight with them.)

After your player has chosen one of the reasons listed, note the letter in parenthesis by it, and light this up on the machine by holding the flashlight behind it. Flash the light on and off and rattle the noisemaker if you have one.

Have your player jot down the letter the machine flashes and, under it, the time-rating (1 to 6) he gave the activity. Then have him look at the other activities on his list. Does he engage in any of them for the same reason as the first? If so, have him compute a time-rating for each and list the number(s) under the same letter. For the remaining activities on his list, have him repeat the whole process. At the end he will have three columns—P, A, and C—each with numbers under it.

Here is an example to make this clear:

Joan has chosen to start with *Cheerleading*. She practices her motions and cheers whenever she gets a chance. Her father has put a mattress in the den so she can practice tumbling. 6 points doesn't seem right, but cheerleading is her main interest right now, and she feels right about rating it 5 points. As to her reason for cheerleading, there's no question—the pure joy of doing it! The machine records "C."

Joan looks over her other activities and decides she started *Dancing* for the same reason. She gives herself 3 points for this—she doesn't spend as much time with it. Now her score is: *C*

5

3.

Joan looks at her *Tennis* item. She enjoys playing it, and is on the courts from the beginning of the season until the end. A 6 for time spent doesn't

100

seem unreasonable. However, Joan doesn't feel the same about tennis as she does about cheerleading and dancing. Right from the beginning she had a special aptitude for tennis and a drive to excel in it. Having won every contest she has entered, Joan is even toying with the idea of becoming a professional. Joan asks to hear the options again. The first one, about its being logical doesn't sound too bad. But number 6—the idea that tennis was teaching her something that would help her later in life—seems better. Joan decides on that one and the machine responds with an "A." *Declamatory* was somewhat the same. She gives herself 2 for time spent on it.

Debating is her next item; she gives it a 4 for time spent. She isn't sure why she got involved in debating. Her father had been very good in it and had talked about it a lot. He felt it had helped him in his profession as a lawyer. So she supposed it must be option two or four—either something she ought to do, or something her parents liked. She decides on number four. The machine flashes "P" for this. *Band* seems similar, so she puts it under P and gives it 4 for time spent. For *Choir*, her next item, she considers her reasons and picks option two. The machine flashes "P"; so she puts it under P, giving it a 3 for time.

Most of the people Joan goes with are crazy about *Skiing*. She doesn't like it as well as other activities, but it is fun to be with the gang. So she picks option five for it and the machine flashes a "C"; she writes down 3 under C for time spent. She likes being with the gang watching *Football* too, but she enjoys watching the game itself more, so picks option three for it. The machine again flashes "C." She gives it a 2 for time spent, and writes this under C. Joan loves *Movies* and fictional shorts on television and watches whenever there is time—which isn't too often. So she gives this a 2 for time spent, and puts the 2 under the C, since she considers movies like football.

Joan's score now looks something like this:

P	A	C
4	6	5
4	2	3
3		3
		2
		2
11	8	15.

Part 3. Have the participants answer the following questions[6] on the basis of how often they occur—giving from one to six points for each answer. A negative answer will, of course, be no points. Continue to itemize the points under the letter the machine will flash. (If they do not participate in group activities, have them answer in the context of what they do every day.)

1. In the activities you especially like, is everything cut and dried where new ideas are not encouraged? (P)
2. When faced with a problem in your activities do you tend to avoid it or try to pretend it isn't there? (C)
3. Do your classmates accuse you of being too serious and not entering into the fun of the activity? (A)
4. If you are not a leader of a group, do you accept the ideas of the leader without expressing differences when you feel them? (C)
5. Do you keep yourself so busy with projects in the clubs and elsewhere that you have no time for fun, or for just doing nothing? (A)
6. Instead of entering wholeheartedly into projects or activities, do you claim you don't know how when you just don't want to make the effort? (C)
7. If you are one of the leaders of the club, do you find yourself telling your classmates that they "ought to" or "should" do a project the way you want it done? (P)
8. Can you sometimes picture yourself as a machine that is just spitting out data, computerized analysis, or decisions? (A)
9. Do you find yourself expressing your parents' value judgments rather than applying your own thoughts and examination to the subject matter? (P)

After your player has time-rated all of the questions, have him ask the machine for the letter involved. Note the letter in parenthesis after each question and light up this letter on the machine. As your player receives the letter for each question, have him add the time-rated number for each to the columns he has made for Part 2.

Part 4. Bring everyone together for the final stage. Remind your players that this is only a game. It is not meant to be exact; circumstances could alter the answers and lack of honesty could distort them. The aim is only to acquaint the players with the different Ego States, and give a "ball park" estimate of your players' Ego States.

Have the participants count up their scores. Then have them draw circles for the letters P, A, and C, each in proportion to the total score for that letter.[7]

> Look once again at Joan:
> Joan realizes that her "A" is a good size and her "P" certainly isn't small; but her "C" seems to be in balance—or should it be larger for this time in life?
>
> Joan glances over at her friend, Bob. She likes him a lot because he is so much fun. He has a huge circle for "C," not too large a one for "P," but his "A" is nonexistent! Bob has laughingly said he will probably just be a bum

when he gets out of school. It sounded like fun when he kidded about it! She notices now that he is looking at his drawing with some concern.

Her friend Susan has a different picture. Joan had thought of Susan when the question came up about how classmates accuse you of being too serious and not entering into the fun. That certainly fit Susan! Her "A" is very large, and her "P" is even larger. Her "C" is the one that is very small. Susan is looking at her picture with surprise.

Joan notices that Dale's picture doesn't seem to be like him. If he did his figuring in as sloppy a manner as he did that project last week, it's no wonder. But after all it's only a game! Gus and Martha and most of the rest have large "C's" and reasonable-sized "P's." Some of their "A's" are small though. But this is supposed to be a fun time in life, isn't it?

There should be some discussion with this game. The questions in Part 4 are on the Constant Ego State (a small distortion of the normal). If these questions do not apply, players should be able to pass them by with negative answers. If anyone has a guilty or negative feeling about a certain question, ask that person if he can figure out why. Is it like a parent looking over their shoulder telling them they are wasting their time? Is it like a child resenting an implied censor?

Explain to your teenagers they are at a crossroad in life and can start to bring their Ego States into balance. Their child should be allowed to come out and play freely with abandon. They should also be growing more aware of the adult within them who can stop and reassess the past, and, looking at the present objectively, use reasoning powers to plan on a reality basis.

Listing as a Discussion Technique: What Is Maturity?

(Taken from Ann Landers)[8]

1. Maturity is the ability to control anger and settle differences without violence or destruction.

2. Maturity is patience. It is the willingness to pass up immediate pleasure in favor of the long-term gain.

3. Maturity is perseverance, the ability to sweat out a project or a situation in spite of heavy opposition and discouraging setbacks.

4. Maturity is the capacity to face unpleasantness and frustration, discomfort and defeat, without complaint or collapse.

5. Maturity is the ability to make a decision and act. However, action without careful thought can be as bad as inaction. The mature person takes the time to consider realistic possibilities, and consequences.

6. When a decision turns out to be wrong, maturity is being big enough and humble enough to say, "I was wrong," and change. And, when right, the mature person need not experience the satisfaction of saying, "I told you so."

7. Maturity means dependability, keeping one's word, coming through the crisis. The immature are masters of alibi. They are the confused and disorganized. Their lives are a maze of broken promises, former friends, unfinished business, and good intentions that somehow never seem to materialize.

8. Maturity is the art of living in peace in spite of that which we cannot change.

(Maturity is a controversial subject where no two people agree. The reader will have his own ideas; so will the teenagers! Teenagers are constantly coming up against the word "immature." Some discussion on the subject should be beneficial.)

Suggested reading:
Muriel James and Dorothy Jongeward, *Born to Win* (Reading, Mass.: Addison-Wesley Publishing Co., 1971).

6

the love-
marriage-
divorce
syndrome

The need to love and be loved starts the moment a child is born and continues throughout his or her life. There are many types of love and all of them play a part in the life of a teenager. The most troublesome, however, is the mating instinct.

Carl Jung sees "first love" or "love at first sight" as a transference. There is, Jung says, a feminine ideal or image in each man, which he calls an *anima*, and a masculine image in every woman, which he calls an *animus*. Ordinarily, in adolescence we project this image onto a qualified person, and have the earth-shaking experience of "falling in love." When the projection is withdrawn the love is ended. This is commonly called "puppy love" and most young people experience it numerous times. Projection alone may be the reason "behind a lifetime of many loves."[1] Real love, although it may start with projection, becomes deepened and cannot be as easily withdrawn.

Whether it stems from projection or not, however, an unrequited love can cause shattering pain and bitterness. Roleplaying can bring emotional release; and the discussion following can alert the players to the fact that almost everyone goes through the pangs of unrequited love but grows in understanding from it.

EARLY MARRIAGE

Roleplay 13

*Jean: Jean realizes that her teenage marriage has not been working out the way she thought it would. Her husband has been spending more and more time with the boys. She is unhappy and disillusioned. This is why she agreed when he said he wanted a divorce. But what happened? Is there something wrong with her? She *had* tried.

Lester: Lester realizes that he should never have married so young. He had been so much in love that he had needed to have Jean belong to him. Now all that feeling is gone and he just doesn't like being married. He would rather be with the boys. She is a nice girl and he hates to hurt her feelings. He feels guilty about that. However, he is simply not in love with her any more. Come what may, he wants out!

[As the scene opens Jean and Lester are discussing a trip she was to make home. She had thought a separation might be an answer. She now realizes it would make no difference.]

Jean: Be honest with me, Lester. You'd be glad to see me go and just stay away, wouldn't you?

Lester: I guess our marriage isn't working out so well.

Jean: Is it the responsibility? I have a job now and I'll keep working. We don't need to have any children unless you want them.

Lester: It's not that, Jean. You've been wonderful. It's not your fault.

Jean: But then why? Why?

Lester: I don't know, Jean. I don't like being married. I like being with the boys, and I don't like feeling guilty. I just don't know why the marriage hasn't worked out for me.

*Role descriptions and dialogue are for the conductor only; they are given as guides.

Suggestions for the Roleplay "Early Marriage"

Selecting Your Players. Nothing special here. The girl and the boy should be of equal fighting ability. If one or the other must be a lightweight player, it should be the role of the boy.

Instructing Your Players. Remember not to tell your players what to say. You should be trying to give them the facts of the role and some psychological help to put them into the mood of it. To Jean you could say something like: "You wonder what happened to your marriage. Did you do something wrong? Lester had been *so* in love with you. You've tried hard to make it a success. How have you failed? Is there something about *you* that killed his love?" Etc. And to bring her into the scene: "You had thought a short separation might help the marriage—maybe even bring back that wonderful feeling of closeness. Now you are convinced it wouldn't make any difference. But you feel you've got to talk it out. You have to know what happened and there's no better time than right now. You're so terribly lonely!" Etc.

To Lester you could say: "You want out of this marriage—but quick! There's nothing wrong with Jean; she's a great gal. But all the ecstasy of being with her is gone—there just isn't anything left! You would rather be with the boys!" Etc. And to plunge him into the scene: "You're glad Jean is leaving on this trip. After she's gone for a while it might be easier to urge her to seek a divorce. You really don't want to hurt her, but you can't stand living without freedom! And you're lonely besides." Etc.

What to Watch for. This role deals with the boy's projection of his anima, his "puppy love." The players should be leveling with each other and trying to understand what has happened. Watch for the "there must be something wrong with me" feeling in the player of the girl-part. This has been built into the role. How is she handling it? Is she retaliating with put-downs? The role-dialogue given shows Jean on the defensive but not reacting in this way. It also shows leveling and how it brings the players quickly and directly to the problem they are facing. They should be trying to arrive at some solution that does not leave the girl-player with a feeling of rejection or the boy with a feeling of guilt. Does either player try to deal with the idea (given in the instructions) that they are feeling lonely? Do they try to answer why she is unhappy or he feels there is nothing left to the marriage? This part of the role is trying to get at maturity and the lack of intimacy between the couple.

Discussion Following the Roleplay. After the role is finished both players should be asked how they are feeling. If they feel rejected or guilty the roleplaying was not a complete success. Have each player state what he or she would like to have heard from the other player. Living together without marriage should be discussed and whether or not it would make any difference in how they would feel when they split up. If the conductor feels it will help, he can have one cast play the role as if married and another cast as if merely living together. The need for maturity when living together or planning marriage should be brought up and discussed. Discuss marriage itself. Its pros and cons. Discuss commitments and why business partners find it advantageous to have a written agreement.

Listing as a discussion technique can be used to advantage here. At the end of this chapter (p. 125) is the discussion question: "What is love?" "What is maturity?" (p. 104) is also especially appropriate for this roleplay. Also at the end of this chapter are games to emphasize the points brought up (pp. 120-124).

(The roleplays in this book are written to stimulate dialogue and communication, and almost anyone can use them. Many of the roles, however, can be changed by someone with counseling experience to meet the needs of more serious problems. This role, for example, could be strengthened to deal with serious rejection by changing the instructions to Jean: delete any mention of divorce. When Lester asks for a divorce in the roleplay it should come as a surprise to her. A person with counseling experience could also add other problems he would like to deal with.)

BROKEN PLANS FOR MARRIAGE

Roleplay 14

*Clay: Clay has been engaged since his junior year in high school. Janet, whom he loves deeply, has not only broken the engagement several times but keeps on postponing the wedding. Two months ago they set a date which was to have been next week. He had sent out invitations to all his friends. Now she has phoned him long distance telling him she will not marry him. He has tried to do everything he could for her, even giving up golfing which she did not like. He must have her. Life just won't be worth anything without her!

Janet: Janet, who is a graduating senior, isn't sure why she doesn't want to marry Clay. He would be a good provider and he will do almost anything she asks him to. There have been times when she felt she was in love with him. But this has only been when she was with him; afterwards she realizes something is wrong. Is it because she just does not love him enough? Or is it something else? How will she make Clay accept the fact that she will not marry him? Last time he sent out invitations to his friends to force her to set a date. This time she is sure and determined!

[Scene opens when Janet comes to answer the door late at night and finds Clay there. Clay has flown home not long after getting Janet's call. He has walked out on his performance at the ballet.]

Janet: Clay, you promised to finish what you had to do in New York before coming home. Did you leave before the performance?

Clay: Aren't you going to ask me in?

Janet: Come in. [She walks into the room and sits down. He follows her and also sits down.] How do you think the new Director is going to take your leaving?

Clay: I really don't care. I can't perform unless everything is settled between us. I'm no good without you, Janet. I just can't live without you!

Suggestions for the Roleplay "Broken Plans for Marriage"

Selecting Your Players. Either the players should be equal in ability or greater weight should be given to the boy-player. An aggressive salesman-type boy would work well if you wish to highlight a person who gives in under pressure.

*Role-descriptions and dialogue are for the conductor only; they are given as guides.

Instructing Your Players. Try to help each of your players believe in the role he'll be playing and identify with it. To Clay you could say something like: "You are very deeply in love with Janet. Why does she keep torturing you by putting off the wedding? Come hell or high water you're going to make sure she marries you this time. These postponements are murder! They're upsetting your job and everything else." Etc. And to bring Clay into the scene: "At the door of Janet's house you hesitate. How are you going to convince Janet to stop this nonsense? You have the license—maybe you can convince her to elope with you tonight. How can she be moved in that direction? By the light jolly fun of an adventure? By proving your deep love in some way? Would she be influenced if you told her you'd commit suicide if she didn't marry you tonight?" Be creative! You are trying to get your player to manipulate.

To Janet you could say: "You don't know why you don't want to marry Clay, he seems everything a husband ought to be. But something is wrong, you have a gut-level feeling that shouts, 'No!' You don't want to hurt his feelings, but if you marry him you'll regret it for the rest of your life." Etc. And to bring her into the scene: "That night you open the door and there stands Clay. He must have missed his performance."

What to Watch for. The emphasis in this role is twofold: on the person who gives in although it is against a subconscious feeling to do so, and on the one who fights for what he wants even if he hurts everyone including himself. Watch for the player who will give in rather than bruise another's feelings. Notice where leveling is used and where it is not. The Janet-player should try to understand her own feelings against the marriage and level with Clay about them. (Reasons should come from player's experience.) Watch for what the players forget or ignore in the instructions. If the boy-player is the hard-sell-salesman type, watch for defensive actions on the part of the girl, or her inability to fight this type of manipulation. The boy-player is not playing a prop role. The role may trigger a feeling of rejection in him, changing the emphasis of the role.

Discussion Following the Roleplay. Discuss gut-level feelings. What are they? Is it wise to ignore them? How can you bring these feelings to the surface so that you can understand them? If two casts are used, which player handles the Janet-part better? Discuss the type of manipulation written into the boy-player's role. Why is it unfair? Why does it endanger happiness? How can the boy win, yet be the big loser? Discuss the aims of good marriages and why elements written into this role go against them. A prudent boy-player may choose to ignore some of the instructions and play the role as his conscience dictates. This is legitimate and should be discussed. Be careful not to condemn a player who in spite of how he felt used the instructions given.

THE NEED TO BE LOVED AND NEEDED

Roleplay 15

*Mother: Comes home from a bridge party and finds a note from her husband, Steve, saying he has a dinner appointment. She had a few cocktails at the card party so she just continues to drink. She wonders how her son Brien is. Are they teaching him self-sufficiency at the military academy? Does her boy need her now? Was her husband right when he said her love for the boy was ruining him?

Father: Coming home from a successful business appointment, he notices the lights are still on but hopes his wife, Ida, has gone to bed. He is too tired to deal with her foibles tonight. She certainly ought to have been able to find something to amuse herself—God knows he gives her enough money! Always fussing about the boy. Well, he misses their son as much as she does. And if she loves the boy as much as she says, she ought to be glad to do what's best for him. Her drinking is beginning to be a real problem.

Brien: He has run away from the academy because he has to know *why* he is being sent away to school. Is he so obnoxious that his parents can't even stand the sight of him? He'd shape up if his mother would only tell him what was wrong. If they loved him wouldn't their actions show it? As he enters the house, he notices his mother has been drinking so he slips up to his room. It would be better to talk to her in the morning.

[As Steve, the father, comes in the door, his wife comes to greet him with a drink in her hand. Brien comes down the stairs when he hears their voices.]

Mother: I'm glad you finally got home. I stayed up to have a talk with you.

Father: We'll talk in the morning. You sound as if you've had a little too much to drink.

Mother: I've had a lot to drink but I'm not drunk. I know because I still hurt.

Brien: (coming down the stairs) You're not the only one who's hurt, Mother. Neither one of you answered my letter about coming home.

Father: Are you in trouble?

Brien: No. I just have to know why I was sent away.

---

*Role-descriptions and dialogue are for the conductor only; they are given as guides.

Suggestions for the Roleplay "The Need to Be Loved and Needed"

Selecting Your Players. Your players should be of equal fighting ability. The part of Brien can of course be played by a girl. Say that she is being sent away to school even though a good high school is available near home. She could be called Rhoda.

Instructing Your Players. Try to give each of your players confidence in his role. But also try to heighten the conflict. After giving Brien, the boy-player, some of the facts, you might add something like: "You don't know why you pulled some of those dumb things, but it wasn't so that they would send you away. Can you make them understand that you don't know why you do these things? Can you tell them you need their love and understanding?" Etc. And to bring him into the scene: "After reaching your room, you hear both your father and your mother talking. You decide you might as well get this over tonight."

To the mother-player you could say something like: "Isn't it a mother's role to love a child? Besides, you have to compensate for a father who isn't sufficiently interested to spend time with the boy. A boy needs his father around—expecially for discipline." Etc. And to plunge her into the scene: "You hear the car, and then the door slam, and you go to the front hallway where your husband is taking off his coat. You've had just enough to drink to talk up to him!"

To the father-player you could say: "You earn the money—the very least she could do is amuse herself! Does she have any idea of the pressures you're under? You wish you had a tenth of the time off she wasted. You would have spent it with the boy. *You* could have given him love without smothering him! But she can't even discipline herself, much less the boy." Etc. And to bring him into the scene: "As you take off your coat and hat, your wife comes to greet you with a drink in her hand. How long has she been drinking? You are too tired to fight with her, so how can you talk her into going to bed?"

What to Watch for. The emphasis of this role is mainly on the boy-player's problem, so he should be sent in almost as soon as the parent-players start talking. The boy's need for love is built into the role. Young people often express their need for love by means of unacceptable behavior—if nothing else, it's sure to get attention! It's a manipulative technique, however, and young people are often unaware that they're doing it, so it is built into this role to bring it to attention. Is the boy-player aware that the parent-players also have problems? Is he trying to help in a constructive way?

The emphasis of the role is on love. Are the parent-players showing the boy-player love? Words alone may not be enough. The act of sending the child away to school says a lot more to him than their words may. How do the players overcome this? Watch for the way things are said. Do "body language" and tone of voice express any deep feeling? Do they convey concern? Is the boy-player, who is supposed to be receiving the love, actually feeling it? (This can be brought out in the discussion afterward, or the conductor can stop the playing and ask the boy-player how he is feeling.) The parent-players should try to show love and consideration toward each other as well as the boy.

The roles of the parents are more than prop roles. The youths playing them will tend to imitate their own parents, and will probably act this way later in real life. Watch for the player who takes the role of father as a "heavy" and makes little attempt to justify it. The mother-player may also see her role as an alcoholic without rights. This should be brought up in the discussion; ask members of the audience to tell how they would handle it. Two casts could also be comapred.

Discussion Following the Roleplay. The discussion with this role could center on love. Was each of the players *feeling* love? Were they given the feeling of being needed? (If love is not being felt, the conductor may want to play some of the practice games for showing love and concern given at the end of the chapter.) Did each of the players give as much as he could to the role? Was any player "snowed under" by the strong offense or defense of another player? Point out where leveling helped or where a lack of it slowed the action down. There might be a discussion on how non-verbal messages sometimes carry more weight than what is being said with words.

DID IT HAVE TO BE DIVORCE?

Roleplay 16

*Ben Boyd: Teenage boy whose parents have problems that might lead to divorce. His mother is a sensitive and understanding person but doesn't apply this to her husband. His father, although he means well, is not sensitive. At a summer job with his father's company, Ben had learned that his father is an excellent executive, but holds an exacting position in which there is constant pressure. This morning his father had asked him if he had taken any liquor out of the cabinet. He told him he hadn't, but afterwards wondered if he should have taken the blame. Mom was going to spring her new job on Dad tonight. That would be rough. Can he help in any way? Is he somehow to blame for their problems?

Ivy Boyd: Because of her husband's lack of consideration in front of people, it has been increasingly hard for Ivy to entertain or go to company parties. If it weren't for her university courses (which he considers harmless) she would have given up long ago. She now has a Ph.D. in bacteriology and has been offered a good job with a firm that knows and respects her work. She needs this, for her husband has made her feel unwanted and inadequate. But taking this position might cause a divorce. She debated whether she had the right to deprive five children of their father even if he didn't understand them. Finally, she accepted the position and now intends to fill it, come what may. In frustration she had accidentally smashed a number of bottles of liquor.

Guy Boyd: Guy is an executive in a large company. He wonders how he could have been so unlucky as to pick a wife with so little capacity to grow. Ivy's always talking about a job—what a laugh! She's so inadequate! Divorce might have been an answer, but he feels the children need him. She would certainly make a mess of their lives! He has noticed that she's been drinking more lately, but he was amazed when he checked the liquor cabinet. Is she an alcoholic?

[The scene opens after dinner when the children have taken off to various parts of the large home and the servants are waiting to clear the table. As they leave the dining room and enter the living room, Ivy turns to Guy.]

---

*Role-descriptions and dialogue are for the conductor only; they are given as guides.

Ivy: I'd like to talk to you about something, Guy. [He shrugs his shoulders and they sit down.] Once, when I wanted to work you embarrassed me in front of an employer. I have accepted a position with Kendal laboratories and I advise you not to interfere.

Guy: I give you plenty of money. You don't *need* to work. The place for a woman with children is in the home!

Suggestions for the Roleplay "Did It Have to Be Divorce?"

Selecting Your Players. This role has been structured so that teenagers can play all of the parts. Players can be of equal fighting ability, or, if your players are unmatched, the mother-player should be the lightweight and the father-player the heavyweight.

Instructing Your Players. Address your players in your own words; try to instill confidence in them for their roles and try to produce conflict. To the boy player you could say something like: "You love both your parents, and want to be with both of them. Is there anything you can do to prevent the divorce your mother has been talking about? Would it have helped if you had taken the blame for the liquor? You feel you should have been helping your parents in some way." Etc. And to bring him into the scene: "You have come back to the dining room to get your history book and you hear your parents talking in the living room. You stay to listen hoping you can help somehow." Conductor can tell the boy-player when to join the others or leave it up to him.

To the mother-player you could say: "You've put up with Guy tearing down your self-esteem long enough. Hired people take care of the work around the house and you need to feel valuable somewhere. The children are all in school and you can arrange your work so that you will be at home when they are. If Guy interferes in your taking the job in any way, you'll ask for a divorce. Children or no children—you've had it!" And to plunge her into the scene: "As you walk out of the dining room into the living room, you realize that Guy may be taking off shortly, so you might as well get it over with. Waiting won't make it any easier to tell him about the job." You are building in reluctance which a shy player might pick up.

To Guy, the father-player, you could say: "You've been forced to entertain your clients at a motel because your wife has refused to be a hostess. It's humiliating! You've put up with Ivy in a lot of things because of the children, but putting up with her as an alcoholic would be too much. You'd get custody of the children—they certainly didn't need a mother like that! Besides, you support a large family with a very demanding job, and you need a little under-standing! You're lonely!" Etc. And to bring him into the scene: "You follow Ivy into the living room. You have a meeting later but first you've got to find out about that liquor." The father-role as cast is that of a dominant, opinionated person, but it will depend on the player as to how it is played.

What to Watch for. How does the boy try to help his parents? Does he try to take their problem on himself (or "own" their problem)? This is written into the role. Watch to see if this happens. Watch for both good and bad ways the boy tries to help. Perhaps he will try to help by taking the blame for the liquor. This may cause complications. Two people can usually work out a conflict more easily without a third person. Is the boy sensitive enough to understand this? Does he slow the action or help it? If the roleplay stagnates because of him, the conductor can send in another player to tell him he is needed elsewhere or call him to the phone.

Are all the players listening sensitively enough to be aware of the feelings behind the words of the others? Is there a hidden message they aren't hearing? Watch for leveling about the facts each player has received. Are they learning from each other by careful questioning and listening? The parent roles are more than prop roles, since young people tend to handle a problem the way they have seen their parents handle it. This is the Parent Ego State of the P A C concept of Berne and Harris explained above (p. 99). Unless they use new knowledge and experiences to alter it, young people will simply repeat this way of doing things, and pass it on to their children in turn.

Discussion Following the Roleplay. At the end of the roleplay allow each player to express how he is feeling. If two casts are used, compare one with the other. If the boy-player takes the blame for the liquor does he help or hinder the conflict between the parents? Discuss the complications of a third party's presence in the conflict. Does it help or hinder? What actions of the boy-player helped the situation most? Which of his actions were of least value? Did the boy-player get the idea that *he* was the cause of the parents' quarreling? Discuss this feeling and why it is normal for children to feel that way.

Were the parents able to get the liquor problem brought out into the open? Was the question of the wife's working settled? While handling their own problem, how are the parent-players treating the boy? Discuss marriage and the need of support from a marriage partner.

NEIGHBORS

Roleplay 17

*Neighbor: Miss Teresa Lambert is an elderly lady who never married. It is hard for her to get along with her next-door neighbors. The children there have very bad manners—they've stomped on her flowers, ruined her trees with climbing, and are now even chasing her cat, Mickie. They keep putting their caged parakeet out on the backyard picnic table even though she had warned them. If Mickie knocked it over and gobbled up the parakeet, there would be the dickens to pay. At the risk of being called a fussy old maid, she is determined to have it out with them.

Mother: Gay has tried to get along with her neighbors, but the elderly woman next door is impossible. The old spinster just doesn't understand children or their needs. She also has a very odd way of looking at problems only from her own point of view. Like the time she wanted them to move their plants because the embankment she had put up so poorly might fall down and crush them. Gay felt that was Teresa's problem not hers, but Roy, her husband, had insisted she move them. Gay is still burned up about that!

Father: Roy can't understand why his wife and children aren't getting along with the sweet little old lady next door. It is about time his children learned a little more consideration for the elderly! As father and head of the household, he's going to see this is corrected. Perhaps he should also insist that his wife show a little more tact.

Anita: Anita, who is eight years old, hates the old hag next door who is so fussy about everything. And Miss Lambert's cat, "dear little Mickie," has been using Anita's sandbox as a toilet! Isn't there a law of some kind that's supposed to keep animals out of other people's yards? The old bag sure seems to think there's a law to keep children out of her yard!

Karl: Karl is a thirteen-year-old boy who tries to get along with the elderly woman next door. But the old witch sure gives him a pain! What did she *expect* him to do when his ball fell into her old flower bed? And her precious trees! Climbing on them doesn't hurt them!

Karen: Karen, who is sixteen, doesn't like the lady next door. Are her parents going to insist she keep her parakeet inside because of the crabby old maid's cat? Dad is unreasonable where this neighbor is concerned. He made her and

*Role-descriptions and dialogue are for conductor only.

117

her mother move their plants to a new location when it was obvious that the dirt embankment Miss Lambert had put up wouldn't hold. The ole biddy just had to suggest it—and Dad agreed. He should have forced her to put up a proper embankment or replace anything it ruined!

[Father is just leaving for work and the whole family is out in the yard. Miss Lambert sees that they are all together and comes out to talk to them.]

Teresa: I thought you would like to know one of your children has left the parakeet sitting out on the picnic table again.

Karen: Mom, you know my parakeet will just die of a broken heart if he isn't allowed outside!

Mother: It's in a cage, Miss Lambert. Perhaps we can just leave it out and watch it very carefully.

Teresa: Then I suppose you will *all* be chasing Mickie, as your little daughter has been.

Father: Anita, I'm surprised at you. You know better than that!

Anita: But Dad, he uses my sandbox as a toilet!

Father: Anita! Did you hear me? No excuses!

Suggestions for the Roleplay "Neighbors"

Selecting Your Players. Players of any fighting ability can be used with this role. This is a group role, so a player who needs experience in how and when to break into a conversation would be given a chance to try out different ways.

Instructing Your Players. Give each player some psychological help to fight for his particular view. Inform the father-player that he will be playing a rigid prop role, called an assignment role. He will side with the neighbor in everything, and show a complete lack of sympathy for his own family. He will be playing the authoritarian role of the prejudiced parent of the P A C concept (explained in the PAC pinball game, page 99). He will hold rigidly to these instructions throughout the playing. Give him a little help psychologically with something like: "You're ashamed of the way your family is reacting to the elderly neighbor next door. Do your children need a little discipline to behave with respect toward her? And your wife is no help. You'd think she'd have the common decency to show a little tact." Etc. And to bring him into the feel of the scene: "Here comes the sweet little old lady now. How can you help her to feel your sympathy?"

To Gay, the mother, you could say: "You wonder if the elderly woman next door will ever understand your children. Or anyone else's problems but her own! Like that time Roy made you move your plants because she was building an embankment that might fall down. She should have been told that if the embankment did fall *she* would have to have them replanted. You still burn when you think about that!" And to bring her into the scene: "Here she comes now with some new complaint. May the Lord give you patience!"

To Teresa, the neighbor you could say: "You've taken about as much as you can from your neighbor's children." Etc. And to bring her into the scene: "The whole family is out in the yard now. It would be a perfect time to approach them, especially since the parakeet is out on the picnic table again. You put down your broom and go over to talk to them."

The children: To *Karen* you could say: "You wish they had to have cats on a leash when they got out of their own yards. Your neighbor acts like she owns your yard too." Etc. To bring her into the scene: "Here comes the old crab now with something on her mind." Etc. To *Karl* you could say: "The old bag next door sure gave you a bad time about her flowers and trees." Etc. To make him feel a part of the scene: "Here she comes now and you thought she hadn't seen you rescue that ball." To *Anita* you could say: "That stupid cat, 'dear little Mickie,' is ruining your sandbox using it as a toilet. Next time you catch it in your sandbox, you're going to do more than just chase it!" And to bring her into the scene: "Oh boy, here she comes now. You're going to let her know she has to keep her cat in her *own* yard. She expects you to stay out of her yard, doesn't she?"

What to Watch for. How are the other members of the family taking the role of the father-player? Are they hurt by his lack of sympathy? How do they handle this? Does the father-player's action inhibit anyone from taking part in the roleplay? Does the authoritarianism and lack of sympahty from the father bring out the protective feeling in the mother-player? Or does the mother-player feel she must back up the father-player? Do the players believe in their roles or is their body language expressing something different? Are the players leveling about the facts of their roles and about their own feelings? Watch for put-downs and manipulations. Watch for sensitive listening or some form of active listening.

Discussion Following the Roleplay. Reveal that the father was a prop player, and then give each player in turn a chance to tell his reaction to the father's authoritarianism and lack of sympathy. If the mother-player has been protective to counteract the lack of sympathy in the father, can she tell *why* she reacted as she did? If, instead, she backed up her husband, did she do it because "she thought she should"? If she did this, was it clear to the others how she really felt? Discuss this concept, remembering that Gordon in his book *P.E.T.* believes both parents should give their honest reaction rather than agree. Did any of the players try to manipulate? If so, discuss the reaction of the player opposing him. Was it hard for him to fight the manipulation?

FUZZY BASEBALL

Game 7

Origin of the word "Fuzzy." The word is taken from a modern fairy tale by Claude Steiner.[2] Briefly, it runs something like this:

"Once upon a time there lived a couple with two children. They were very happy because in those days the good fairy gave everyone at birth a small, soft, Fuzzy Bag containing a supply of Fuzzies that made people feel warm and fuzzy all over. People who didn't get them would develop a sickness and shrivel up and die. Of course they were given freely and very much in demand.

"But the wicked fairy was jealous and she devised a plan to stop this. She told the man that his wife was giving away all of her Fuzzies and soon would not have any left for him. He was very alarmed so he pressured his wife into giving out her Fuzzies more discriminately. The children noticed this and they became stingy. Stinginess spread like wildfire and soon some people were dying.

"The wicked fairy was not interested in having people die, so she gave them a bag with Cold Pricklies that made people feel cold and prickly but prevented them from dying. People began trying to work to earn Fuzzies, but somehow these weren't the same as ones freely given. One entrepreneur even invented the all-purpose Plastic Fuzzy which was supposed to make you feel good but didn't.

"Then a hip woman born under the sign of Aquarius came to the unhappy land, and not knowing about the wicked fairy, again passed out warm Fuzzies to the children. The children loved her and the Fuzzies. They began to give them out freely even though the grownups passed laws against it. What happened then? The outcome is still undecided."

AIM OF THE GAME: To practice giving Fuzzies and get over inhibitions against touching people.

The game challenges players to see the good points in their fellow players. Implies Fuzzies are important.

(This is a game to be played with a light touch.)

PLAYERS: Five to nine players who form a circle.

Conductor will act as Umpire and should be generous in calling the hits: the purpose is to stimulate not judge.

Play Action of the Game:

The player at bat goes to each youth in the circle, giving each one a Fuzzy and touching him or her in some way. Players win bases and bring home players as in baseball. Greeting all the players (or 3 outs) completes the game for each. Player with the highest score wins.

Terminology and Rules for Fuzzies:

What is a Fuzzy?—A Fuzzy is more than a compliment. It does not attempt to evaluate the character of a person, but describes that person's efforts, accomplishments, and the warm, friendly, positive feelings he or she engenders in the sender. For a stranger it might be just something nice you have observed. A Fuzzy is not flattery, it must have some basis of truth.

One-Base Hit—A barely passable Fuzzy. Given only when it is obvious the player is not trying very hard.

Two-Base Hit—Good Fuzzy or good gesture. Good for two bases.

Three-Base Hit—A very good Fuzzy or a very good gesture. If both Fuzzy and gesture are very good, it's a home run.

Home Run—An excellent Fuzzy delivered in a smooth way so as not to make a shy person embarrassed. Or, an excellent or original gesture.
Home run brings in all runners on bases.

Strikeout—Called only if player will not even try.

Caught Fly Ball—If the receiver feels the Fuzzy is flattery, he (or she) can declare it false.
If the Umpire agrees he will declare it a fly ball that was caught and call batter out.
If the Umpire does not agree he will call it an error.

Error—Some people find it hard to accept the truth or legitimate Fuzzies. If the receiver declares false something the Umpire feels is a Fuzzy, he can call the receiver on it, getting the support of the other players.
The Umpire will then declare it an error and the player issuing the Fuzzy will get his hit credited.

Foul Ball—Foul ball is called only if the player is constantly using flattery.
If the Umpire feels a Fuzzy is flattery but the recipient says nothing (Umpire must be sure of this), he can say it looks as if the recipient is a little uneasy.
The Umpire will then declare the Fuzzy borderline or a foul ball and the batter will be given another try.

Terminology and Rules for Gestures (Where different):

Gestures—These can be made in a variety of ways and will depend somewhat on how well the sender knows the receiver and how they feel about each other. It can be a kiss, a vigorous hug, a grasping or a touch or pat on the shoulder, a handshake while looking directly at the person and showing some interest, a shove or a jab, or even a gentle fist to a jaw. The more original the better. (Gestures are judged along with the Fuzzies.)

Caught Fly Ball—If the receiver feels the gesture of the batter was not justified, she can object and call it false. Examples:
   a. A youth who is going with a girl may take her in his arms and give her a gentle and lingering kiss. It may be worth a home run.
   b. But if a stranger (or a boy who is aware that the girl does not like him) takes advantage of the opportunity to act in the same way, the girl can object.

If the Umpire feels the girl is justified, he will call it a caught fly ball and declare the batter out.

If he does not agree he will call it an error.

Error—Some people have grown up without showing or receiving physical affection or touching and may object to any gesture. If a receiver calls a gesture false but the Umpire decides it was genuine, he may declare it an error and the batter will get his man on base.

Foul Ball—If a lightweight player shakes hands in a cold or uninterested way, the gesture could be called a foul ball and the batter encouraged to try again. (Conductor should use this sparingly and only when he can help the player.)

If a player keeps giving only handshakes, he can be warned to try something different by calling a foul on him. If he continues, call it bunting.

Bunting—Gestures that are too much the same. They will put a man on base, but not allow the player on third to get home.

Examples of Fuzzies:
1. "My day is brightened by your cheerful smile."
2. "I envy the way you put words together."
3. "I'm put at ease by your friendly manner."
4. "I like the way your eyes sparkle with interest when I am talking with you."
5. "I enjoy talking with you—you seem to listen with your whole being."
6. "Although I don't know you well, I feel you have confidence in yourself by the way you walk and talk."

FARMER-IN-THE-DELL

Game 8

AIM: To learn to give out friendliness, sympathy, and empathy. To allow others to release hidden resentments or emotion.

PLAYERS: Ten to twelve players who form a circle.

TUNE: Game is played to the tune of "The Farmer in the Dell." ("The farmer in the dell, the farmer in the dell, hi ho the dairy oh, the farmer in the dell.")

Play Action:

Farmer: While the group sings the first stanza of "The Farmer in the Dell," the conductor picks a Farmer to start the game.

The conductor explains that Mr. Dell is a stranger in the community and lonely. He needs the friendliness of his neighbors. As Mr. Dell goes around the circle, each person will say a few words in a friendly way to help him feel at ease.

Wife: While the group sings the second stanza of the song, "The farmer takes a wife," Mr. Dell chooses a wife from the circle.

The conductor explains that instead of choosing a Wife from his new neighborhood, as some of the mothers had hoped, Mr. Dell came back from a trip with his new Wife, who is a divorcee and black. (This is an all-white community.) As Mr. Dell takes his Wife around the circle, each one of his neighbors will say something to help her feel she is one of the group.

Child: Mr. Dell goes back into the circle. As the group sings the third stanza of the song, "The wife takes a child," the Wife will choose a Child from the circle.

The conductor explains that this is a Child from the former marriage of the Wife. Some time ago, the Child was hurt in an accident and has lost the use of his (or her) arms.

The doctors feel it is psychosomatic and the Child just needs a great deal of love.

The Wife will take her Child to each of her neighbors and they will say something, or use a gesture of some kind, that will make the Child feel loved.

Dog: The Wife goes back into the circle. As the group sings the fourth stanza, "The child takes a dog," the Child chooses a player from the circle to be the Dog.

The conductor explains that the Dog is feeling low and depressed. He gets only orders barked at him and no love. He feels unloved and unneeded in a cruel world. The Child will take his Dog to each one in the circle and they will try to cheer him up, make him feel wanted and needed, and show him their love.

Cat: The Child goes back into the circle. While the group sings the fifth stanza, "The dog takes a cat," the Dog chooses a Cat from the circle.

The conductor explains that although all of the characters are essentially prop players to stimulate the reaction of the others in the circle, the next three will be playing nasty characters that make if difficult for the rest to respond. (The conductor may prefer to ask for volunteers from here on, or select someone he knows needs to release resentment or emotion.)

The conductor explains that the Cat is playing the part of a person who has been deeply hurt many times and has learned to have little trust and respect for other people. She (he) makes sarcastic and "catty" remarks. (Player may be helped by picturing a cat with an arched back who spits and scratches when anyone gets near.)

The Dog will take the Cat around the group. No matter what the Cat says to them, the players will not get angry and will always act in a positive and loving way.

Rat: The Dog goes back into the circle. While the group sings the sixth stanza, "The cat takes a rat," the Cat will choose a Rat.

The conductor explains that the Rat is playing the part of a person so deeply hurt that he seems to hate everyone around him. He is, however, reacting to the way he has been treated. He is desperately in need of love, but has learned not to show it by acting in a mean and nasty way.

The Cat will take the Rat around the circle and the players will give back loving words for the Rat's nasty remarks. Gestures may also be used. It's to be hoped the Rat will find it harder and harder to play his part.

Cheese: The Cat goes back into the circle. While the group sings the last stanza of the song, "The cheese stands alone," the Rat will choose a Cheese and then go back into the circle.

The conductor explains that the Cheese plays the part of a person who acts like the big cheese. The know-it-all, the person who thinks he should be leading everything. He is obnoxious in his sense of superiority, riding over other people's comments.

The Cheese will go around the circle by himself—he, of course, doesn't feel he needs any help! Each player will try to tell the cheese what they feel he is doing, but in a loving way. They will say things that will show their love and concern while they are telling him they would like to see him change.

Listing as a Discussion Technique: What Is Love?

After any of the roleplays in this chapter, your teenagers may want to discuss love further. Listing the ideas they have about love will help. Ask them for qualities or definitions of love, and for different kinds. Accept all answers your teenagers give, encouraging them to level about their feelings. (The answers below were given by teenagers.[3] They are merely to stimulate your thinking; do not give them unless absolutely necessary.)

1. Love is a sense of understanding, and of being understood.
2. Love is a relationship in which an individual can say what he feels and know the other person goes right on caring about him.
3. Love is a sense of closeness, of belongingness.
4. Love sees the best in another person, but recognizes the everyday side of him.
5. Love is feeling pain when another person is hurt, or anger when he has been treated unfairly.
6. Love is a feeling of happiness so deep that the person in love wants to make everyone else happy, too.
7. Love is wanting the best possible for the other person.
8. Love expresses itself by making an individual want to be the best kind of person he can be, for the sake of the other.

What Are Different Kinds of Love?
1. Mating love—two people facing life's problems together.
2. Physical love or sexual love.
3. Mother love—a nurturing type of love that gives of itself and protects.
4. Possessive love—a smothering type of love that will not allow another person to be himself.
5. Agape love—love of all God's creatures.

For additional reading, see:
George R. Bach and Peter Wyden, *The Intimate Enemy* (New York: Avon Books, 1970).

conclusion

The time in life known as the teenage is filled with a raft of problems. Yet, oddly enough, it is also a time of almost unlimited possibilities. Youths are very susceptible and impressionable, open to idealism and to the setting of idealistic goals. They reach out (sometimes in a halting or clumsy way) for maturity and adulthood. What an opportunity to lend a helping hand!

I hope this book will help your teenagers, and you, to communicate more effectively. Listening, understanding, building a strong self-image, finding out how to deal with courtship and friendship situations—these are all essential to full communication and growth. I sincerely believe that roleplaying is one of the best ways to learn. It brings out communication problems in a way that the players will never forget, and it will help them for a long time to come because of the experiences they'll have of working out difficulties. Although you can use these roleplays over and over, as you become more adept you may want to produce new roles, using the ones here as a basis and bringing in your own counseling experiences. Or you may want to seek out other sources of roleplay to vary the ones here. If the demand is great enough, there might be a sequel to this book with more roleplays.

Gather together about twenty youths. Take them anywhere—to an isolated cabin in the woods or to a cozy room in a church. It won't matter where you are, for once the roleplaying is started, the world outside fades away. You are dealing with in-depth relationships and nothing is more fascinating or more productive of growth. Good luck!

notes

Chapter 2: Instructions

1. I gained my knowledge of roleplaying techniques in Robert Blees' Training Workshops. You may obtain information on these workshops by writing:

    Robert A. Blees
    2255 Greenville Drive
    West Covina, California 91790.

2. George R. Bach and Peter Wyden, *The Intimate Enemy* (New York: Avon Books, 1970), p. 80.

3. *Ibid.*, p. 324.

4. "We speak at . . . about 125 words per minute . . . . We can think at the rate of 400 to 500 words per minute." From a pamphlet by Robert Haakenson, "The Art of Listening" (Philadelphia: Smith, Kline & French Laboratories, Speakers Bureau & Speech Training Service).

5. From *Are You Listening?* by Ralph G. Nichols and Leonard A. Stevens. Copyright 1957 by McGraw-Hill Book Company. Used with permission of McGraw-Hill Book Company. From pp. 134-137.

6. Thomas Gordon, *P. E. T., Parent Effectiveness Training* (New York: Peter H. Wyden, Inc., 1970), pp. 115-138.

7. *Ibid.*, pp. 121-129.

8. *Ibid.*, p. 63.

Chapter 3: Listening

1. Gordon, *P. E. T.*, pp. 49-61.

2. Copyright © 1970 by Dr. Thomas Gordon. From the book PARENT EFFEC-TIVENESS TRAINING, published by Peter H. Wyden, a division of David McKay Co., Inc. Reprinted & paraphrased with the permission of the publisher. From pp. 41-44.

3. Nichols and Stevens, *Are You Listening?*, chapter 9.

4. *Ibid.*, chapter 11. See p. 18 of the present book.

5. Haim G. Ginott, *Between Parent and Teenager* (New York: The Macmillan Co., 1969), p. 55.

Chapter 4: Understanding

1. Paul Tournier, *To Understand Each Other* (Richmond: John Knox Press, 1967), p. 28.

2. Edward Strecker and Kenneth E. Appel, *Discovering Ourselves* (New York: The Macmillan Co., 1962), pp. 155-161.

3. Frieda Fordham, *An Introduction to Jung's Psychology* (Baltimore: Penguin Books, 1966), pp. 29, 33.

4. Strecker and Appel, *Discovering Ourselves*, pp. 161-169.

5. Tournier, *To Understand Each Other*, p. 31.

6. Maria F. Mahoney, *The Meaning in Dreams and Dreaming* (New York: Citadel Press, 1966), pp. 91-93.

7. *Ibid.*, pp. 93-95.

8. *Ibid.*, p. 89.

9. Fordham, *Introduction to Jung*, pp. 36-38; P. W. Martin, *Experiment in Depth* (New York: Pantheon Books, 1955), p. 22.

10. Fordham, *Introduction to Jung*, p. 42.

11. Jolande Jacobi, *The Psychology of C. G. Jung* (New Haven: Yale University Press, 1962), p. 12.

12. Wheel from *The Psychology of C. G. Jung* by Jolande Jacobi, p. 16. Copyright 1962 by Yale University Press. Used by permission. Sources for adaptation: Martin, *Experiment in Depth*, pp. 22-28; Fordham, *Introduction to Jung*, pp. 36-44.

13. Everett L. Shostrom, *Man, the Manipulator* (Nashville: Abingdon Press, 1967), p. 220. Shostrom says that "one becomes actualizing at that moment when he fully surrenders to the awareness of his manipulations" (p. 222). He believes this occurs in three stages:

    1) Figure out the ways you manipulate and the reason you do.

    2) Restore inner balance by:

       a) Exaggerating manipulative tendencies to experience the foolishness of them.

       b) Expressing the opposite polarity of this pattern.

    3) Put both poles into a unified whole (pp. 218-220).

14. *Ibid.*, pp. 33, 52.

15. *Ibid.*, pp. 36-39.

16. *Ibid.*, figures 1 and 2 on pages 37 and 55. Used by permission.

17. *Ibid.*, pp. 50-51.

Chapter 5: Search for Identity

1. Carl G. Jung, Marie-Louise von Franz, Joseph L. Henderson, Jolande Jacobi, and Aniela Jaffe, *Man and his Symbols* (New York: Dell Publishing Co., 1968), pp. 103-107.

2. Changing the sex of the name can sometimes be accomplished by just adding a feminine or masculine ending. (Feminine endings: a, ia, i, ie, e, ne, ine, isa, itta, itte, ette, it, ica, ika, itsa, iki, anne, illa, alee. Masculine endings: s, as, us, ston, ton, ten, nett, sin, son, ald, dy, le, ley, and, ick, win, ris, ren.) Respelling sometimes involves dropping silent letters (Nicholas, Nichole) and end letters like "w" or "y" (Andrew, Andrea), or changing one letter to another like "y" to "i" or "k" to "c" (Anthony, Antonia). If it becomes necessary to create a new name for a special purpose not mentioned, you can create the name like an *anagram*, switching the letters of a special word or group of words until it has a pleasant sound; *telescoping* by dropping letters from a group of words until you arrive at a suitable name; or *inversion* by switching syllables. From Sue Browder *The New Age Baby Name Book* (New York: Workman Publishing Co., 1974), pp. 16-19.

3. Muriel James and Dorothy Jongeward, *Born to Win* (Reading, Mass.: Addison-Wesley Publishing Co., 1971), pp. 101, 108.

4. Eric Berne, *The Structure and Dynamics of Organizations and Groups* (Philadelphia: J. B. Lippincott, 1963), p. 137.

5. James and Jongeward, *Born to Win*, pp. 127, 128.

6. *Ibid.*, pp. 254-256.

7. *Ibid.*, p. 256.

8. Ann Landers, "Maturity: It Can Mean Many Things," *Wisconsin State Journal* (December 31, 1974), section 2, page 2. Copyright 1974 by Field Newspaper Syndicate. Used by permission.

Chapter 6: The Love-Marriage-Divorce Syndrome

1. Mahoney, *Meaning in Dreams*, pp. 125, 127, 131.

2. Claude Steiner, *Scripts People Live* (New York: Bantam Books, 1975), p. 127. Copyright©1974 by Grove Press. Used by permission.

3. From the book COUNSELING WITH TEENAGERS by Robert Blees. © 1965 by Prentice-Hall, Inc. Published by Prentice-Hall, Inc., Englewood Cliffs, New Jersey. The excerpt is from the Fortress Press reprint (Philadelphia, 1968), p. 115.

8777